Andreas Tomanek

SILICONES & INDUSTRY

A compendium for practical use, instruction and reference.

Published by Wacker-Chemie GmbH, Munich

HANSER

Andreas Tomanek

SILICONES & INDUSTRY

A compendium for practical use, instruction and reference.

About the author

Dr. A. Tomanek was born on 17th September 1937 in the Polish city of Lodz. He studied chemistry at the University of Frankfurt and for the last 20 years has held various posts as a silicone chemist at Wacker-Chemie.

Title page

The title page shows a two-dimensional representation of an NMR spectrum. Actual spectroscopic photographs of polysiloxanes have been taken by the Consortium für elektrochemische Industrie GmbH, our research establishment in Munich.

The molecular structure depicted on the title page is that of a silicone polymer.

Tomanek, Andreas
(Silicone & Technik, English)
Silicones & Industry: A compendium for practical use, instruction and reference/Andreas Tomanek
p. cm.
Translation of: Silicone & Technik. 1990
Includes bibliographical references and index.
ISBN 3-446-17264-5
1. Silicones. I. Title, II. Title: Silicones and Industry
TP 248.S5T6613, 1992   92-16110
62G. 1'93--dc 20   CIP

Publisher: Wacker-Chemie GmbH, Munich
Overall design: Maximilian Heigl, Puchheim
Illustrations: Maximilian Heigl, Puchheim; Pro Publishing GmbH, Munich
Title page: C´Company Werbeagentur GmbH, Munich
Setting and printing: Druck + Werbung Holzmann, Bad Wörishofen
Coordination: Maria Schneider
Distribution: Carl Hanser Verlag, Munich
© by Wacker-Chemie GmbH, Munich 7/1991

This book is protected by copyright. All rights reserved, especially those pertaining to translation, reprinting, reproduction of diagrams, broadcasting, photomechanical reproduction, and storage in data-processing systems.

Translated from the German by Raymond A. Brown

Foreword

Silicones not only play key roles in the latest advances made in space travel, medicine and electronics. They are also an integral part of our daily lives.

Nevertheless, the extent to which they help us solve difficult problems is still not generally appreciated.

The aim of this compendium is both to highlight their unusual properties and to act as a source of inspiration for those with problems of their own to solve.

In addition to providing a general, comprehensible picture of silicone chemistry, the author addresses the problems of the most important industries in which silicones are successfully employed. Silicones themselves therefore act as a vehicle for the reader to become better acquainted with many branches of industry.

"Silicones & Industry" is the outcome of many years of practical experience in the field of silicones and is thus an excellent textbook and reference work.

Wacker-Chemie GmbH, Munich

Table of contents

	Page
1. Silicon in nature	9
2. The chemistry of silicon	10
3. The structure of silicones	12
4. Direct synthesis of silanes	13
5. Special silane syntheses	18
5.1 Addition reactions (hydrosilylation)	18
5.2 Recombination and disproportionation reactions	18
5.3 Nucleophilic substitution reactions	19
5.4 Grignard syntheses	20
6. Hydrolysis to polysiloxanes	20
6.1 Manufacture of silicone fluids and rubber polymers	22
6.2 Manufacture of silicone resins	24
6.3 Manufacture of silicates	25
6.4 Fumed silica	26
7. Industrial silicone products	27
8. General properties of silicones	28
8.1 Resistance to heat and cold	28
8.2 Dielectric properties	30
8.3 Hydrophobic properties and release action	30
8.4 Surface activity	32
8.5 Physiological properties and environmental aspects	32
9. Silicone fluids and derivatives	33
9.1 Properties of silicone fluids	33
9.2 Silicone copolymer fluids	35
9.3 Functional silicone fluids	37
9.4 Silicone defoamers and foam stabilizers	37
9.5 Silicone fluid emulsions	40
9.6 Release agents	42
10. Silicone rubber	42
10.1 Properties of silicone rubber	49
10.2 Processing of RTV silicone rubber	63
10.3 Processing of HTV silicone rubber	66
11. Silicone resins	71
11.1 Properties and types of silicone resins	71
11.2 Processing of silicone resins	72
12. Chemicals, petrochemicals and coal industries	74
12.1 Recovery and refining of crude oil	74
12.2 Chemical feedstocks	78
12.3 Polymerization	79
12.4 Pharmaceutical products	81

	Page
12.5 Man-made fibres	85
12.6 Treatment of pigments and fillers	89
12.7 Chemical consumer products	92
13. Plastics industry	94
13.1 Modification of plastics	94
13.2 Auxiliaries for plastics processing	97
14. Rubber industry	99
14.1 Articles made of silicone rubber	99
14.2 Silicone auxiliaries for the rubber industry	100
15. Textiles and leather	103
15.1 Silicone finishes	104
15.2 Dyeing and printing	106
15.3 Garments industry	108
16. Paper industry	108
16.1 Manufacture of paper	108
16.2 Silicone release papers	111
16.3 Printing techniques	114
17. Construction industry	116
17.1 Joint sealants	116
17.2 Masonry protection	123
17.3 Insulating materials	132
18. Surface coating compounds	133
18.1 Heat-resistant paints	136
18.2 Coatings with greater weathering resistance and flexibility	137
18.3 Anticorrosion coatings	138
18.4 Coatings additives	138
19. Electrical and electronics industries	139
19.1 Cables and cable accessories	139
19.2 Electrical insulation materials	143
19.3 Transformers	147
19.4 Electric heat and lighting technology	147
19.5 Office automation	148
19.6 Entertainment electronics	150
19.7 Electronic components	151
20. Metals, machines, ceramics	153
21. Transportation	155
22. Foodstuffs, pharmaceuticals and medicine	158
Terms used in chemistry and industry	162
Index	168
Literature	172

1. Silicon in nature

Silicon is the most abundant element in nature after oxygen and makes up more than 25 % of the earth's crust.

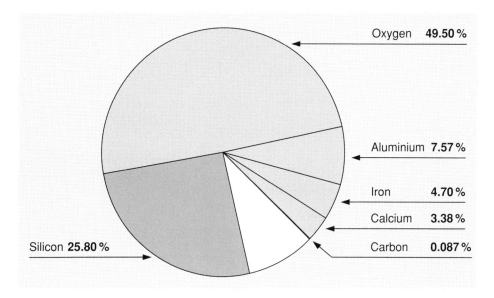

On account of its great affinity for oxygen, silicon – unlike carbon – is found not in the pure form but mainly in compounds of silica, especially the silicates of magnesium, calcium, iron and mica. It also occurs in the form of sand, quartz and pebbles. A common feature of all these compounds is their tetrahedral structure in which the silicon atom is bonded to four oxygen atoms.

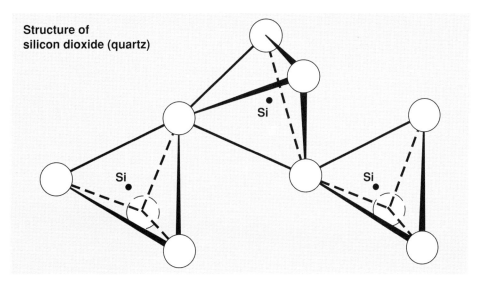

Structure of silicon dioxide (quartz)

Silicon is the paraplastic substance of many microorganisms in the animal kingdom, e.g. radiolarians, infusorians and diatoms. Their remains are familiar to us today as kieselguhr and diatomaceous earth. Silicon is also absorbed in small amounts by higher life forms, including humans. However, it cannot enter the respiration and assimilation cycles enjoyed by (monomeric) carbon dioxide because it forms polymeric Si – O compounds.

Although the ancients made extensive use of siliceous compounds such as sand, clay and ceramics for construction and working materials, it was not until 1824 that elemental, amorphous silicon was first isolated by Berzelius. Nowadays, three forms of silicon predominate in industry, namely

– ferrosilicon and calcium silicide ($CaSi_2$) : for making alloys in the steel industry,

– technical grade silicon (assay 98.5 – 99.7 %) : for the aluminium and chemicals industries,

– hyperpure silicon: for the semiconductor industry.

Silicon carbide (SiC) is an important industrial abrasive, and is used for refractory brick and high-temperature furnaces. Recently, it has found application as a high-performance material for valves, sliding bearings etc.

Silicon compounds are much rarer in nature than their carbon counterparts and not even synthesis methods are capable of closing the gap in the respective chemistries.

The first synthetic silicon compounds date back to the years between 1904 and 1940 when Kipping was working extensively on the synthesis of organochlorosilanes. He produced compounds that were unparalleled in nature and showed that many interesting syntheses are feasible. However, the fact remains that the chemistry of silicon simply cannot rival that of carbon.

The significance of polymeric silicon compounds was recognized at a late stage and it was not until Rochow presented his direct synthesis in 1940 that they became economically viable. The first major applications for them were developed in subsequent years. In the U.S.A., Hyde adapted them for electrical insulation while, in the U.S.S.R., Andrianov conducted fundamental research work into organosilicon compounds. In Europe, the first company to manufacture silicones was Wacker-Chemie, largely as a result of the work performed by Nitzsche.

2. The chemistry of silicon

Silicon is located in Group 4A of the periodic table and exhibits both metallic and non-metallic characteristics. Its properties are thus very similar to those of carbon but major differences do exist, the primary causes of which are given below.

– Silicon is much more electropositive than carbon.

C	H	Si
2.5	2.1	1.8
C^-H^+		Si^+H^-

- The silicon atom is larger than the carbon atom.

C	Si
0.77 Å	1.15 Å

- The electron configurations of both elements are:

$$1s^2\,2s^2\,2p^2 \qquad\qquad 1s^2\,2s^2\,2p^6\,3s^2\,3p^2$$

- With oxygen, which is more electronegative, silicon forms very stable single bonds, whereas carbon prefers double bonds.

$$\underset{}{H_3C-\overset{\overset{O}{\|}}{C}-CH_3} \qquad\qquad H_3C-\underset{\underset{OH}{|}}{\overset{\overset{OH}{|}}{Si}}-CH_3$$

- Silicon is very reluctant to enter into double bonds, and indeed forms only a few, unstable silane compounds. The reason for this lies in the much lower energy required for $\pi \rightarrow \pi^*$ transitions in alkenes and disilanes, as is evident from the corresponding molecular orbital diagrams ($2p\pi \rightarrow 2p\pi^*$ for C is approx. 6 eV; $3p\pi \rightarrow 3p\pi^*$ for Si is approx. 3 eV).

- Compounds of silicon and carbon are mainly tetravalent. However, silicon is also capable of higher and lower co-ordination.

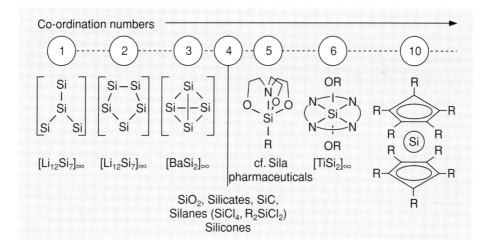

Unsaturated molecules in co-ordination numbers 1, 2 and 3 have been synthesized in the gas phase and are encountered in metal silicides. Many complexes have co-ordination numbers 5 and 6. Higher co-ordination numbers have recently been discovered in sandwich complexes of silicon with cyclopentadiene ligands and in others with dicarborane ligands.

3. The structure of silicones

Silicones are essentially quartz-like structures in which the three-dimensional SiO_2 backbone has been modified by incorporation of methyl groups. This progressive saturation leads ultimately to low polymers.

Linear silicone polymers usually have the following structure.

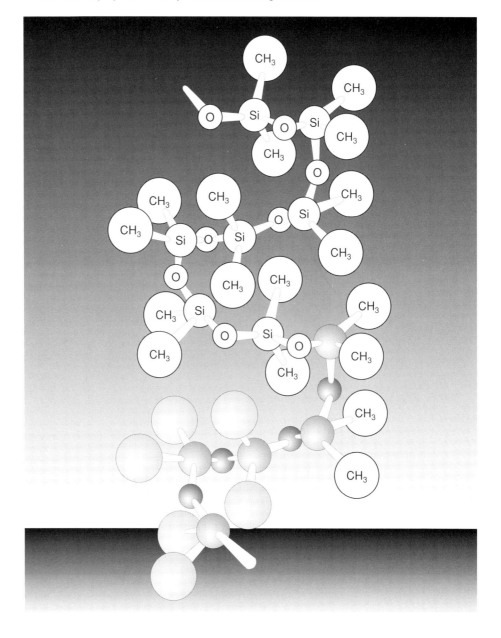

The polymers are typically tangled, with the methyl groups able to rotate freely about the Si – O – Si chain. Silicones are also referred to as polysiloxanes, which is their correct chemical name.

Silicone chemists distinguish siloxanes by their functionality.

R | R – Si – O – | R	R | – O – Si – O – | R	| O | – O – Si – O – | R	| O | – O – Si – O – | O |
M mono- functional	**D** di- functional	**T** tri- functional	**Q** quadri- functional

Monofunctional units (symbol M) act as chain terminators. When they undergo hydrolysis, the simplest molecule formed is $(CH_3)_3 Si – O – Si (CH_3)_3$.

Difunctional units (D) form the backbone of higher molecular chains and cyclic compounds.

Trifunctional (T) and quadrifunctional (Q) units form space-network molecules. The backbone of quartz, for example, is composed exclusively of Q-units.

Functionalities can be interconverted by means of chemical reactions. For example, T-units can be rendered into Q-units:

$$H – Si \lessgtr \xrightarrow{\text{Alkali}} KO – Si \lessgtr$$

The use of diverse organic groups considerably extends the scope for variation. Methyl, phenyl, vinyl and several higher alkyl groups are among those of economic importance.

4. Direct synthesis of silanes

In industry, almost all methyl chlorosilanes are made from silicon and monochloromethane (methyl chloride) by the Rochow-Müller direct synthesis. The raw materials for silicones are thus sand and petroleum (shale). The reaction is carried out at a temperature of 250 – 300°C in the presence of copper catalyst.

$$Si + 2CH_3Cl \longrightarrow (CH_3)_2SiCl_2 \text{ and other silanes}$$

The purity and composition of the silicon and the copper catalyst are crucial to the smoothness of the reaction. For industrial manufacture, the silicon must have an assay of around 97 % and a particle size of 45 – 250 µm. Greater purity (> 99 %) results in a slow reaction while a lower content (< 95 %) often leads to poor yields and an unfavourable crude silane distribution.

Best results are obtained with a mixture of silicon and a small percentage of copper. A high proportion of copper (20 – 80 %) initiates the reaction very rapidly but leads to a poor conversion rate because the silicon is consumed too quickly. The Rochow synthesis is extremely sensitive to certain metallic poisons. These must not be present in concentrations greater than specified limits, such as 0.2 - 0.5 % for aluminium and several ppm for lead, because otherwise the reaction does not proceed as desired.

The reaction can be controlled by adding up to 0.2 % zinc dust, $ZnCl_2$, or $ZnSO_4$ in order to

– accelerate it at lower temperatures, i.e. enhance activity, and

– obtain a high proportion of $(CH_3)_2SiCl_2$, i.e. increase relative yield.

The extent to which, for example, 50 ppm of lead can influence relative yield and the conversion rate for silicon is shown in the following table.

Influence of lead on the direct synthesis method

Silane	Percentage yield	
	< 5 ppm Pb	≥ 50 ppm Pb
$(CH_3)_2SiCl_2$	83	60
CH_3SiCl_3	10	20
$(CH_3)_3SiCl$	3	3
CH_3SiHCl_2	4	17
Si conversion (%)	≈ 100	≈ 50

Although the reaction is severely inhibited by lead, it is promoted by judicious amounts of antimony.

Gases can also be helpful (e.g. HCl, H_2), serving above all to enhance relative yields. With their aid, it is possible to render the silanes halogen-rich, halogen-deficient or hydrogen-rich.

A catalyst of finely ground silicon and copper is used for the main reaction, although comminuted silicon-copper alloy can serve as an alternative. The reaction is thought to proceed via the formation of an intermediate.

$$\text{Si}---\text{Cu} + \text{CH}_3\text{Cl} \longrightarrow \begin{array}{c}\text{Si}---\text{Cu}\\ |\quad\quad |\\ \text{CH}_3---\text{Cl}\end{array} \longrightarrow \text{CH}_3-\overset{\backslash/}{\text{Si}}-\text{Cl} + \text{Cu}--$$

Direct synthesis is strongly exothermic and requires precise **temperature control** because a high space-time yield and high yields of the principal product, dimethyl dichlorosilane, are only obtained when the temperature lies between 250°C and 300°C. The temperature is regulated by means of a reactor cooling jacket or the addition of an inert gas.

Above all, it is essential to thoroughly mix the Si – Cu catalyst and the CH_3Cl in a fluidized bed. Nowadays most silanes are synthesized industrially in this way.

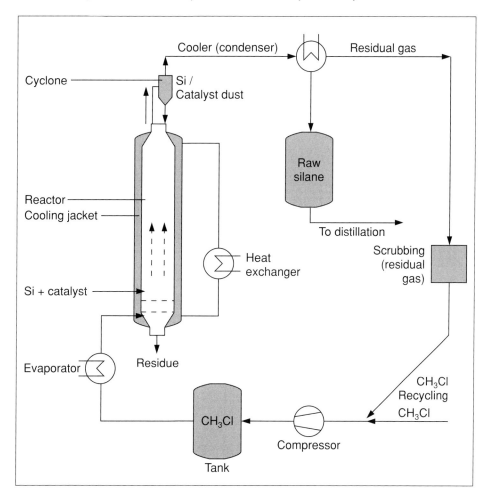

The bed of silicon and copper in the reactor is impinged upon by a powerful stream of monochloromethane introduced tangentially under pressure. The strong turbulence is created in order to prevent hot spots from forming. The temperature in the reactor is 280°C. If it is lower, conversion is too sluggish and, if it is higher than 320°C, cracking occurs. The reaction proceeds best at a pressure of 1 – 5 bar. Any solid particles that form are separated off. The gas phase is condensed to raw silane liquid and gaseous monochloromethane, which is recycled.

Modern fluidized bed reactors are capable of producing approximately 40,000 tonnes of raw silane annually, the usual composition of which is as follows.

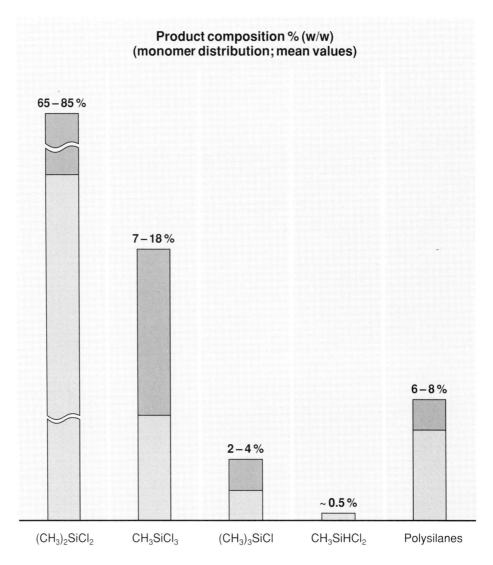

The raw silane mixture is then separated into its constituents by a series of **distillation** columns.

The dimethyl dichlorosilane must have a high assay because even a small amount of a trifunctional or monofunctional silane can cause chain branching (content of CH_3SiCl_3 must be < 0.2 %) or inhibition (in the case of $(CH_3)_3SiCl$). Since the boiling points of CH_3SiCl_3 and $(CH_3)_2SiCl_2$ differ by a mere 4°C, fine distillation must be performed to obtain the dimethyl dichlorosilane. Assays yielded by the distillation processes are as follows.

1) Distillation of the raw silane mixture: 99.8 %
2) Fine distillation: 99.99 %

Monomethyl dichlorosilane (CH_3HSiCl_2) must also be more than 99 % pure. Another problem encountered is the formation of azeotropic mixtures of $(CH_3)_3SiCl$ and $SiCl_4$.

The distillation columns must have up to 200 trays and a reflux ratio of 1 : 500. For reasons of safety, they are cooled with air and not water.

Name	Formula	Boiling point
Dimethyl hydrogenmonochlorosilane	$(CH_3)_2HSiCl$	35°C
Methyl hydrogendichlorosilane	$(CH_3)HSiCl_2$	41°C
Trimethyl chlorosilane	$(CH_3)_3SiCl$	57°C
Methyl trichlorosilane	$(CH_3)SiCl_3$	66°C
Dimethyl dichlorosilane	$(CH_3)_2SiCl_2$	70°C
Trimethyl trichlorodisilane	$(CH_3)_3Si_2Cl_3$	152–156°C
Dimethyl tetrachlorodisilane	$(CH_3)_2Si_2Cl_4$	
Heavy fractions		> 156°C

Non-usable by-products, such as high-boiling methyl chlorosilanes (approximately 3 % of product), can be separated off with the aid of HCl in the presence of an amine catalyst and mostly recycled to dimethyl dichlorosilane.

Methyl chlorosilanes undergo rearrangement on treatment with $AlCl_3$ to ultimately yield $(CH_3)_2SiCl_2$ as well.

$$(CH_3)_3SiCl + CH_3SiCl_3 \xrightarrow{AlCl_3} 2(CH_3)_2SiCl_2$$

Extra scrubbers can be installed after the hydrolysis stage. The raw siloxane mixture, consisting of cyclic and linear siloxanes, is converted to cyclosiloxanes and then purified by distillation.

5. Special silane syntheses

Syntheses other than the Rochow synthesis are described below.

5.1. Addition reactions (hydrosilylation)

Silanes and siloxanes that contain a bound hydrogen atom can add across double and triple bonds.

$$RCH = CH_2 + H-Si\lessgtr \longrightarrow RCH_2 - CH_2 - Si\lessgtr$$

The reaction can be made to proceed either by means of free radicals or under the influence of catalysts of noble-metal compounds.

In the latter technique, the co-ordination number of the noble metal, which is usually platinum, alternates continuously, this process being accompanied by adsorption and desorption.

Platinum-catalysed hydrosilylation is used extensively to synthesize organofunctional silanes, i.e. silanes bearing amino groups, vinyl groups etc.

5.2. Recombination and disproportionation reactions

These types of reactions serve mainly to alter the degree of substitution of silanes and thereby render them into industrially useful products.

$$SiR_4 + SiR_4 \underset{\text{Disproportionation}}{\overset{\text{Recombination}}{\rightleftarrows}} 2\,Si(R\,R)_2$$

The reaction is carried out in the presence of a Friedel-Crafts catalyst, e.g. organic amines and aluminium chloride.

5.3. Nucleophilic substitution reactions

Chlorine atoms, hydrogen atoms and alkoxy groups can undergo nucleophilic substitution. A typical example is the synthesis of phenyl silanes:

$$CH_3H\,SiCl_2 + C_6H_5Cl \xrightarrow{550-650°C} C_6H_5(CH_3)SiCl_2 + HCl$$

One advantage is that no toxic, chlorinated biphenyls are formed.

Nucleophilic substitution is widely used as a means of synthesizing functional silanes.

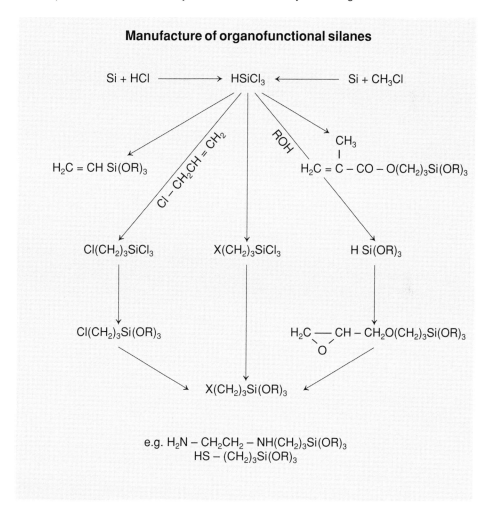

5.4. Grignard syntheses

Grignard reagents can be used to attach organic groups to silicon atoms direct.

$$RMgCl + Cl-Si\lneq \longrightarrow RSi\lneq + MgCl_2$$

This reaction is highly versatile and can be readily controlled. Consequently it is commonly used to prepare organosilicon compounds on a laboratory scale. Grignard reagents are also used to make phenyl silanes.

6. Hydrolysis to polysiloxanes

All organopolysiloxanes are obtained by the **hydrolysis** (or alcoholysis) of chlorosilanes. The resultant silanols immediately undergo polycondensation.

$$(CH_3)_2SiCl_2 + 2H_2O \longrightarrow (CH_3)_2Si(OH)_2 \longrightarrow \text{Polysiloxane}$$

Silane — Silanol

The mechanism of hydrolysis and the composition of the silane mixture govern the size of the polysiloxane molecule, the degree to which it is crosslinked (i.e. its crosslinking density) and the number of hydroxyl groups that it contains.

Both intramolecular and intermolecular condensation can occur, with the former yielding ring structures and the latter, linear polymers. Intramolecular condensation is favoured by excess water, solvents or neutralization of the hydrochloric acid generated.

Polycondensation is all the more likely when the silanols have a low molar mass, several OH groups are attached to the silicon atom and the organic group is small.

Polycondensation of siloxanes

$$(CH_3)_3SiOH + HOSi(CH_3)_3 \xrightarrow[-H_2O]{} (CH_3)_3SiOSi(CH_3)_3$$

$$\underset{\underset{CH_3}{|}}{\overset{\overset{CH_3}{|}}{HO-Si-OH}} + \underset{\underset{CH_3}{|}}{\overset{\overset{CH_3}{|}}{HO-Si-OH}} \xrightarrow[-H_2O]{(+H^+)} \underset{\underset{CH_3}{|}}{\overset{\overset{CH_3}{|}}{HO-Si-O}} - \underset{\underset{CH_3}{|}}{\overset{\overset{CH_3}{|}}{Si-OH}}$$

$$\underset{\underset{CH_3}{|}}{\overset{\overset{CH_3}{|}}{HO-Si-O}} - \underset{\underset{CH_3}{|}}{\overset{\overset{CH_3}{|}}{Si-OH}} + n(CH_3)_2Si(OH)_2 \longrightarrow \underset{\underset{CH_3}{|}}{\overset{\overset{CH_3}{|}}{HO-Si-O}}\left[\underset{\underset{CH_3}{|}}{\overset{\overset{CH_3}{|}}{Si-O}}\right]_n\underset{\underset{CH_3}{|}}{\overset{\overset{CH_3}{|}}{Si-OH}}$$

α, ω –Dihydroxypoly-dimethylsiloxane

$$\xrightarrow[-H_2O]{+(CH_3)_2Si(OH)_2}$$ Octamethyl-cyclotetrasiloxane

Methanolysis has important advantages over hydrolysis as a means of generating polysiloxanes from silanes. One of them is that the HCl formed in the reaction can be converted to monochloromethane and used in the Rochow synthesis.

$$(CH_3)_2SiCl_2 + 2CH_3OH \longrightarrow [(CH_3)_2SiO]_n + 2\,CH_3Cl + H_2O$$

$$Si + 2\,CH_3Cl \longrightarrow (CH_3)_2SiCl_2$$

The process takes place in the gas phase, with neutralization being effected by a continuous infusion of steam. Linear polysiloxanes constitute the main products and are obtained in a yield greater than 98 %. Very small amounts of cyclic compounds are formed.

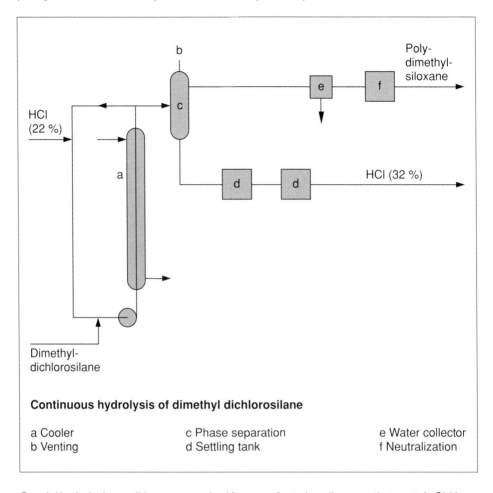

Continuous hydrolysis of dimethyl dichlorosilane

a Cooler
b Venting
c Phase separation
d Settling tank
e Water collector
f Neutralization

Special hydrolysis conditions are required for manufacturing siloxanes that contain Si-H bonds, including the use of toluene and ethanol.

6.1. Manufacture of silicone fluids and rubber polymers

The hydrolysate (α, ω-dihydroxy siloxane whose viscosity is 100 mPa s) is processed further to silicone fluids and rubber polymers.

Processing takes the form of polymerization* in the presence of acidic catalysts ($PNCl_2$, bleaching earths activated with acids) at temperatures of around 130°C. When the reaction is over, the catalyst must be deactivated by neutralization or filtered off.

* Chain stoppers such as $(CH_3)_3Si-O-Si(CH_3)_3$ are used in the case of silicone fluids.

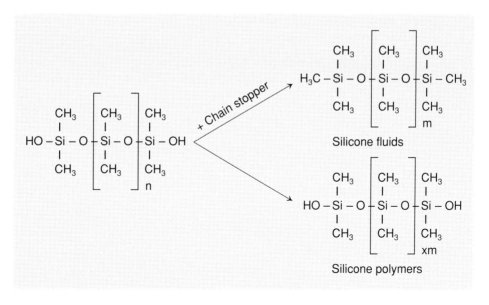

Special demands are imposed on the purity of silicone polymers for manufacturing rubber. For example, the proportion of trifunctional units must be lower than 0.2 %.

Such high purity can be achieved by

— distilling the dimethyl dichlorosilane until it is 99.99 % pure, or

— purifying the dimethyl polysiloxane via cyclic compounds.

In the latter case, the hydrolysate is converted to a mixture of cyclic compounds, which is then purified by distillation.

The conversion is performed at 140°C in the presence of KOH and an inert solvent. Trifunctional siloxanes are bound as silanolates, Si-H groups are destroyed and tetrameric compounds are distilled off. The cyclic compounds are polymerized direct to siloxane polymers.

An equilibration step exploits both the ease with which Si–O–Si linkages are cleaved in the presence of acidic or basic catalysts and the rapidity with which new Si–O–Si bonds form. Chain cleavage occurs concurrently with polycondensation and polymerization. Thus, inhomogeneous mixtures of siloxanes of various chain lengths are converted to polymers whose chain lengths are uniform and follow a gaussian distribution.

6.2. Manufacture of silicone resins

When trichlorosilanes and tetrachlorosilanes undergo hydrolysis, the first chlorine atom is attacked so rapidly that too much crosslinking occurs too quickly and a gel forms.

It is therefore desirable to control the extent of this instantaneous polycondensation in such a way that the siloxanes formed remain soluble.

This can be accomplished only by either batch processing or alkoxylation.

6.2.1. Batch-process hydrolysis

The first step in the batch process is to partially hydrolyse the silane in a mixture of water and organic solvent (e.g. toluene). Since the reaction medium is highly dilute in solvent, hydrolysis cannot go to completion. The silane and siloxane dissolve in the organic solvent, thereby escaping hydrolysis by the water and the acid.

$$R-\underset{\underset{Cl}{|}}{\overset{\overset{Cl}{|}}{Si}}-Cl + H_2O \longrightarrow R-\underset{\underset{Cl}{|}}{\overset{\overset{(OH)_x}{|}}{Si}}-(OH)_x + HCl$$

In the second step, excess water is used to induce large-scale polycondensation.

$$-\overset{|}{\underset{|}{Si}}-OH + HO-\overset{|}{\underset{|}{Si}}- \xrightarrow{H^+} -\overset{|}{\underset{|}{Si}}-O-\overset{|}{\underset{|}{Si}}-$$

6.2.2. Alkoxylation

In this method, the silanes are converted first to intermediates, some of which contain alkoxy (methoxy) groups, and then to the silicone resins.

Silane + methanol / water ⟶

Intermediates

$$\begin{array}{cc} & OCH_3 \\ & | \\ Si & Si \\ | & | \\ OCH_3 & OCH_3 \end{array}$$

Intermediate + water ⟶

$$\begin{array}{cc} OCH_3 & \\ | & \\ Si & Si \\ | & | \\ O & O \\ | & | \\ Si & Si \\ & | \\ & Si \\ & | \\ & OCH_3 \end{array}$$

These intermediates are of especial importance in the coatings industry (e.g. combination resins) and in masonry protection.

Solvent-free encapsulating resins function on the principle of addition reactions between two components, one of which contains hydrogen atoms and the other, vinyl groups.

$$\geq Si-H + CH_2 = CH-Si \leq \xrightarrow{Pt} \geq Si-CH_2-CH_2-Si \leq$$

Trifunctional silanes can be converted to water-soluble siliconates, which turn into insoluble silicone resins only on exposure to atmospheric carbon dioxide.

$$CH_3SiCl_3 \xrightarrow{H_2O} \underset{\text{Methyl silicic acid}}{[(CH_3Si(O_{1/2})_2OH]_x} \longrightarrow \underset{\text{K-methyl siliconate}}{CH_3Si(OH)_2OK}$$

The siliconates are made by running methyl trichlorosilane into constantly circulating concentrated hydrochloric acid.

6.3. Manufacture of silicates

Silicates are made by alkoxylating $SiCl_4$. They contain 40 % SiO_2 and are the raw materials for pressure-sensitive adhesives, PVC-foam stabilizers and stone-strengthening compounds.

6.4. Fumed silica

Pyrolysis of silanes yields fumed (or pyrogenic) silica, which is employed as a filler.

The manufacturing processes for various silicas are shown in the diagram below.

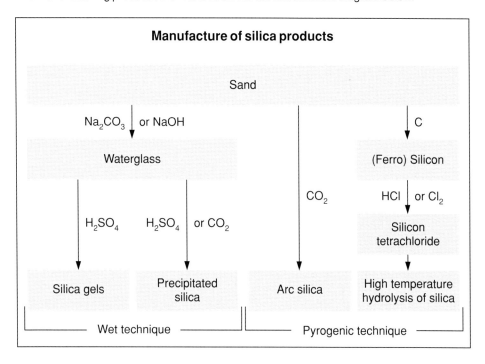

Production of various silicon products, such as silanes, silicones, silicas, can be neatly summarized in a flow chart, such as that given below.

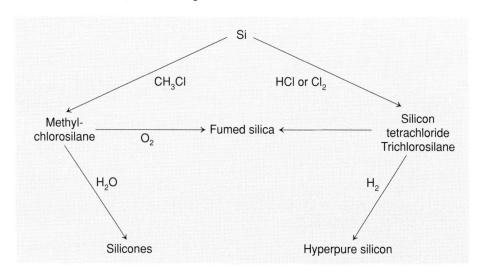

7. Industrial silicone products

Methyl chlorosilanes, alkoxy silanes and silanes containing organofunctional groups are popular industrial surface-impregnation agents, e.g. for fillers, pigments and masonry.

Methyl chlorosilanes are made in by far the greatest quantity and serve primarily as raw materials for silicone polymers. A plethora of industrial silicone products are derived from them.

The relative economic importance of silicone products is shown below.

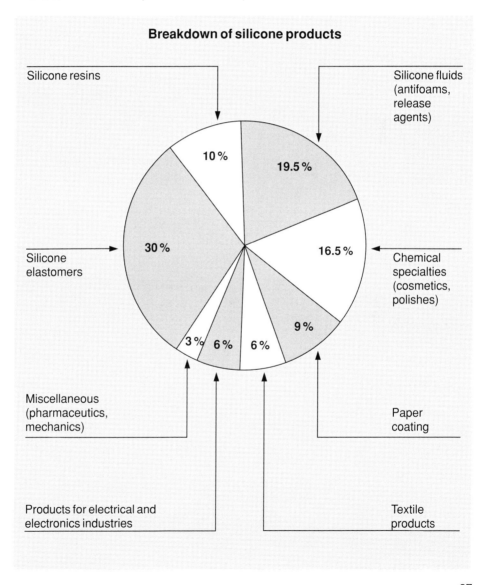

8. General properties of silicones

Silicones are imbued with a number of outstanding properties that render them indispensable for many applications.

8.1. Resistance to heat and cold

The thermal resistance of silicones is such that their physical and mechanical properties alter only very gradually with changes in temperature. Extensive breakdown of properties occurs only at temperatures in the range of 200 – 300°C. Alteration of properties mainly takes place under one of two conditions, namely in the presence or absence of oxygen.

– Atmospheric oxygen primarily oxidizes organic groups. This manifests itself first as embrittlement or gellation and then progressive decomposition with the evolution of volatile products (above all: CO_2, H_2O, HCHO) and formation of silica.

– Absence of atmospheric oxygen leads to depolymerization, which results in the formation of low molecular siloxanes, especially under the catalytic influence of acids and bases.

Comparisons of silicones with other organic plastics resistant to temperatures of up to 250°C must take account of respective processability.

Type of plastic	Processability
Polyimides Polyetherimides Polysulfones	Only possible in solution
Silicones	In solution (aqueous and organic solvents)
	Extrusion and injection moulding
	Solvent-free encapsulating materials

Silicones' high thermostability stems from the strength of the Si – O – Si linkage.

A look at bond energies shows that the Si – O bond, which has a bond angle of 130°, is stronger than the C – O bond.

Bond	Bond energy
Si – O	444 J/mol
C – O	339 J/mol

At the other end of the temperature scale, the products' resistance to low temperatures manifests itself in low pour points and glass-transition temperatures of around -60°C and -120°C respectively.

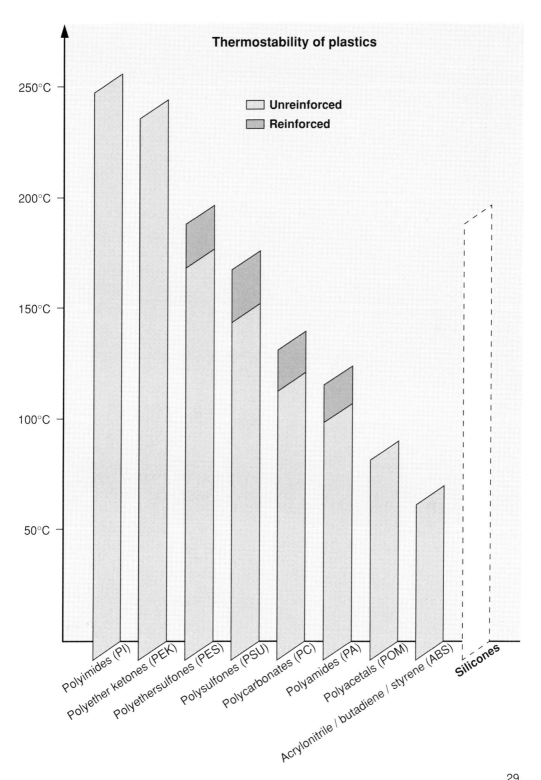

8.2. Dielectric properties

Silicones rank among those plastics with the best dielectric properties, i.e. they are good insulators. Values for dielectric strength, resistivity and dielectric constant do not change much over a wide temperature range (20 – 200°C). In particular, silicones are highly resistant to tracking.

Even when silicones decompose under the action of heat, the product formed is silica, which is itself a good insulator and affords emergency operating properties.

8.3. Hydrophobic properties and release action

Silicones are extremely hydrophobic, a fact which is exemplified by comparing the contact angle against water droplets on various substrates. The greater the contact angle, the greater is the water repellency.

Type of surface	Contact angle against water droplets
Glass (degreased)	0°
Polyethylene	ca. 95°
Paraffins	ca. 105°
Silicones (after baking on glass)	100 – 110°

The strongly hydrophobic nature stems from the methyl groups. Silicone fluids spread out spontaneously over surfaces for which they have an affinity, e.g. glass, building materials. They are also used in the textiles industry to make special mordants (catalysts), such as ZnO_2, and in heat fixing.

A major advantage of silicones over paraffins is that they spread out spontaneously and thinly, with the result that only small amounts are required for imparting the desired level of water repellency. A further bonus is that they interact very little with the surface of the substrate, whose appearance and ability to breathe are retained.

Release action of silicones

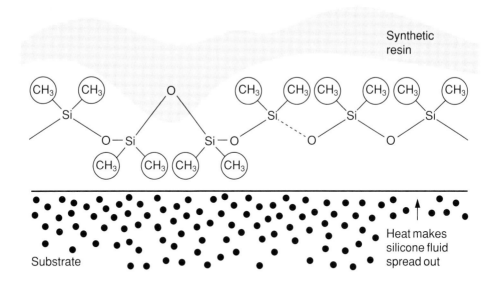

Closely allied with hydrophobicity is the ability of silicones to act as release agents for tacky substances such as fats, oils and plastics.

By contrast, functional silanes promote adhesion.

How a silane adhesion promoter works

8.4. Surface activity

The great ease with which silicone fluids spread out over surfaces stems from low intermolecular forces of attraction. A measure of these forces is afforded by measurements of surface tension. Silicones are found to have low values.

Product	Surface tension
Water	ca. 75 mN/m
Polysiloxanes	ca. 20 mN/m
Surfactants	ca. 30 mN/m

Surface activity is a chief factor in the use of silicones as antifoams, foam stabilizers, and as free-flow agents in paints and flotation aids.

8.5. Physiological properties and environmental aspects

With the exception of the low-viscosity, volatile compounds, silicones are remarkably inert towards living organisms. There are two reasons for this. First, the polymeric nature of the products hampers passage through biological membranes and, second, biological systems have difficulty in breaking the Si-C bond. When taken orally, silicones exert only a laxative effect. The LD_{50} for rats is greater than 30 ml/kg. Studies of sub-chronic and chronic oral toxicity have revealed neither toxic effects nor evidence of carcinogenicity. Silicones have been tested extensively as materials for prosthetic devices.

They enter the environment primarily in the form of fluids and derivatives such as antifoams, waxes, and cosmetics. Experience has not revealed any toxic, mutagenic or teratogenic effects on animals or aquatic life and there is no evidence that silicones adversely affect ecosystems. Silicones degrade on contact with micaceous clays to form cyclic siloxanes. A combination of ultraviolet light and nitrates converts siloxanes to silica, which, unlike silicones, can be absorbed by siliceous algae.

Let us now take a closer look at the properties of the various classes of products and at how the products are processed.

9. Silicone fluids and derivatives

9.1. Properties of silicone fluids

The most important physical properties of silicone fluids are shown below.

Physical properties of silicone fluid of viscosity 350 mPa s	
Molecular weight (number average)	ca. 10 000
Flash point	> 300°C
Solidifying point	- 50°C
Thermostability	≤ 200°C (in air)
Ignition point	around 500°C
Thermal conductivity at 50°C	0.15 W/K/m
Dielectric strength	14 kV/mm
Resistivity	6×10^{15} Ω cm
Surface tension	21 mN/m

A general idea of the thermostability of silicone fluids is provided by the following table.

Temperature	Stability
≤ 200°C	Stable for a number of weeks (viscosity 100 – 1000 mPa s)
250°C	Gelling sets in after 24 h. Stabilized fluids increase in viscosity after a number of weeks (oxidation is delayed by stabilizer.)

A striking feature of silicone fluids is that their physical properties, such as viscosity and thermal conductivity, are largely unaffected by temperature.

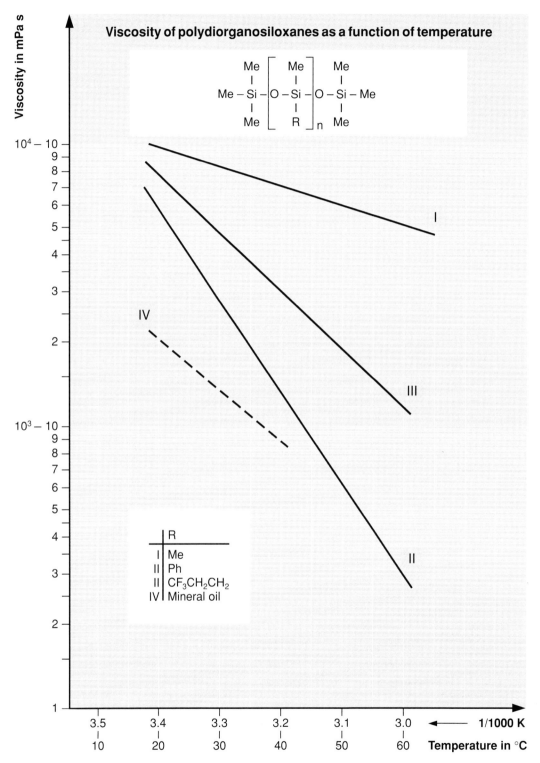

By virtue of their low intermolecular forces, silicone fluids have a liquid consistency over an extremely wide range of molecular weights.

Viscosity (mPa s)	Molecular weight	Average chain length
0.65	162	0
50	3 000	40
100	5 000	70
1 000	15 000	200
10 000	37 000	500
100 000	74 000	1 000

Fluids with low viscosities have solidifying points of around -50°C and those with viscosities of at least 1000 mPa s have very low vapour pressures.

The boiling points of the low-viscosity silicone fluids are lower than those of carbon compounds of similar composition and approximately the same molecular weight.

– $(CH_3)_3Si-O-Si(CH_3)_3$ boils at 101°C

– Isoamyl ether boils at 173°C

The flash points of silicone fluids whose viscosities exceed 100 mPa s lie above 300°C, with spontaneous ignition occurring at temperatures above 420°C. Compressibility is relatively high at $\approx 100 \times 10^{-12}$ cm^2/dyne, being greater than that of mineral oils. An important characteristic is that their viscosity changes much less than that of mineral oils when they are subjected to pressure, particularly over protracted periods of time. For example, after 200 000 pressure cycles within the space of 500 hours, their viscosity would have changed by only 2 % whereas that of mineral oils would have fallen by about 50 %.

Silicone fluids have a thermal conductivity of 0.15 W/K/m, which is equivalent to 0.00037 cal /cm/s/°C (0.001 in the case of special silicone heat-sink pastes). The thermal conductivity values for aluminium, water, glass and air are 0.48, 0.014 and 0.000057.

Silicone fluids can be modified chemically (e.g. copolymerized), emulsified and compounded (reinforced or thickened with fillers) to yield a large number of industrial silicone products, the most important of which are presented below.

9.2. Silicone copolymer fluids

Modification to the siloxane structure takes the form of substituting alkyl chains for methyl groups or effecting copolymerization with organic polymers. For example, incorporation of poly(ethylene oxide), $-CH_2CH_2O-$, or poly(propylene oxide), $-CH_2CH_2CH_2O-$, yields siloxane copolymers, which, unlike pure polysiloxanes, exhibit a more or less strongly hydrophilic nature. This is reflected in water-solubility above a certain cloud point and emulsifying or surfactant properties. Accordingly, such modified siloxanes find use as wetting agents, and as flow improvers in paints etc. Modified siloxanes are made as required by means of hydrosilylation, co-condensation or re-esterification.

Some typical structural groups are shown below.

Polysiloxane – polyether

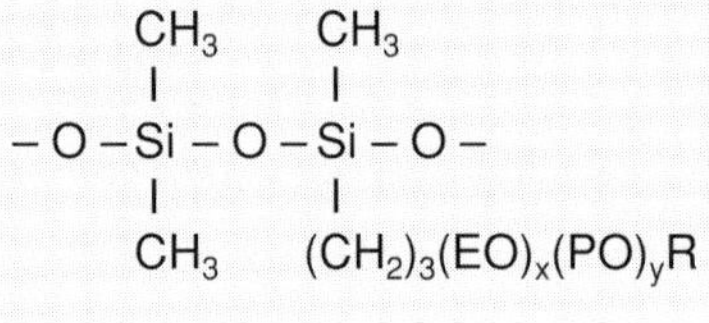

$$
\begin{array}{c}
\quad\ CH_3 \quad\ \ CH_3 \\
\quad\ \ | \quad\quad\ \ | \\
-O-Si-O-Si-O- \\
\quad\ \ | \quad\quad\ \ | \\
\quad\ CH_3 \quad (CH_2)_3(EO)_x(PO)_yR
\end{array}
$$

Polymethylsiloxane – polyalkylsiloxane

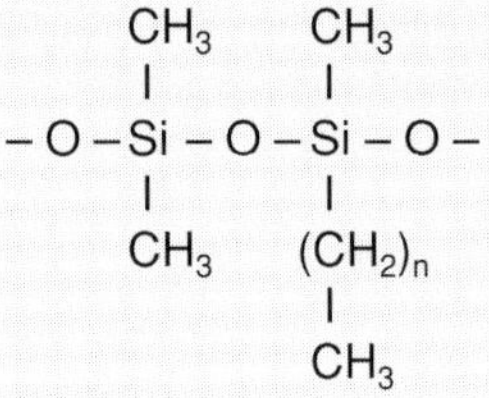

$$
\begin{array}{c}
\quad\ CH_3 \quad\ \ CH_3 \\
\quad\ \ | \quad\quad\ \ | \\
-O-Si-O-Si-O- \\
\quad\ \ | \quad\quad\ \ | \\
\quad\ CH_3 \quad (CH_2)_n \\
\quad\quad\quad\quad\quad\ | \\
\quad\quad\quad\quad\ CH_3
\end{array}
$$

Cationic polysiloxane (silicone surfactant)

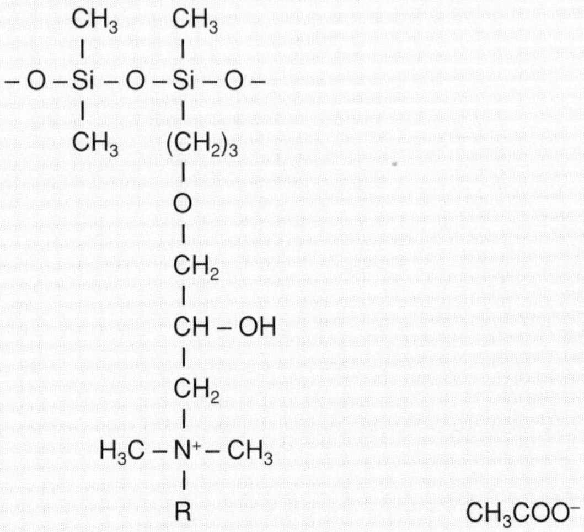

$$
\begin{array}{c}
\quad\ CH_3 \quad\ \ CH_3 \\
\quad\ \ | \quad\quad\ \ | \\
-O-Si-O-Si-O- \\
\quad\ \ | \quad\quad\ \ | \\
\quad\ CH_3 \quad (CH_2)_3 \\
\quad\quad\quad\quad\quad\ | \\
\quad\quad\quad\quad\quad O \\
\quad\quad\quad\quad\quad | \\
\quad\quad\quad\quad\ CH_2 \\
\quad\quad\quad\quad\quad | \\
\quad\quad\quad\quad CH-OH \\
\quad\quad\quad\quad\quad | \\
\quad\quad\quad\quad\ CH_2 \\
\quad\quad\quad\quad\quad | \\
\quad\quad\quad H_3C-N^+-CH_3 \\
\quad\quad\quad\quad\quad | \\
\quad\quad\quad\quad\quad R \quad\quad\quad\quad CH_3COO^-
\end{array}
$$

Polymethylsiloxane – polyalkylsiloxane – polyether

$$\begin{array}{c}
CH_3 CH_3 CH_3 \\
| | | \\
-O-Si-O-Si-O-Si-O- \\
| | | \\
CH_3 (CH_2)_n (CH_2)_3 \\
 | | \\
 CH_3 (EO)_n \\
 | \\
 R
\end{array}$$

Polysiloxane-polyethers can be linked together in a linear, alternating arrangement but branched structures are also feasible. Lower siloxanes possess special structures.

Trimeric, phenylmethyl siloxanes of low molecular weight enjoy great economic importance as diffusion-pump oils. Some derivatives of cyclic siloxanes have a waxy consistency, e.g. tetra(cyclohexa) silane.

Products containing these structural groups are more compatible with mineral oils and waxes, and are used, for example, as release agents. This application demands a certain degree of compatibility with coatings, waxes and adhesives etc.

The siloxanes may also contain both saponifiable (Si – O – C) and nonsaponifiable (Si – C) linkages.

9.3. Functional silicone fluids

Functional silicone fluids are siloxanes with reactive end groups and differ from unreactive methyl silicone fluids in this respect. Examples of the former are OH-polymers (hydrolysate), H-siloxane, and silicone fluids that bear amino or epoxy groups.
They are obtained by hydrolysis of the corresponding chlorosilanes.

$$\geq Si-(CH_2)_3NH(CH)_2NH_2 \longrightarrow \underset{\diagdown\diagup}{Si} - O - \underset{\diagdown\diagup}{Si} - O - \underset{\diagdown\diagup}{Si} - (CH_2)_3NH(CH)_2NH_2$$

(with $H_3C\ CH_3$ groups on each Si)

Functional siloxanes, especially amino-silicone fluids, are characterized by having a high substrate affinity and find application in textile finishing, hair cosmetics, car polishes etc. They are also used in conjunction with H-siloxane for crosslinking various organic plastics.

9.4. Silicone defoamers and foam stabilizers

Silicone fluids have a pronounced antifoaming action and are the basic components of special defoamers. The latter usually contain an activating solid, such as fumed silica.

Defoamers must satisfy some important criteria for effective foam destruction. They must

– be insoluble in foaming media,

– have a very low surface tension, and

– spread out rapidly over foaming media, displacing the foam-forming agents from the surface.

In order to be able to understand the defoaming action, it is necessary to know how foam is generated. Foam occurs when surfactants accumulate in solution at the interface with air (air bubbles), thereby reducing the surface tension of water. Provided that too much surfactant does not drain away, the bubbles remain stable. However, with increasing drainage, the bubbles become thinner and thinner until finally they burst.

Antifoaming agents accelerate this decomposition process in various ways.

Silicone fluids penetrate the bubble film on account of their good spreadability, entraining with them hydrophobic solids, such as fumed silica. Spreadability and foam penetration can be enhanced by resorting to special emulsifying additives known as entering agents. A typical entering agent is butanol.

The mechanisms by which the different defoamers work are many and varied and depend on numerous factors, including the nature and stability of the foam, surface tension and surface viscosity, rate of drainage, healing effect.

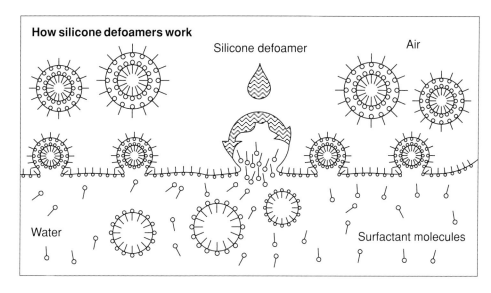

The defoaming mechanisms of the most important classes of products are shown below.

Type of defoamer	Mechanism
Silicones	Reduction in surface stability
Mineral oils	Reduction in surface stability
Fatty acids and glycerol	Reduction in surface stability
Organic alcohols and esters (butanol, glycols)	Healing effect
Salts of organic acids	Change in surface potential
Tributyl phosphate	Drainage

The applications in which silicone defoamers are used are so varied that all kinds of formulations have to be developed to satisfy the different properties required. For example, it may be necessary to employ a hydrophilic or hydrophobic base oil.

The most important criteria to be satisfied are

– resistance to heat and alkalis,
– low tendency to adsorb suspended particles, e.g. during filtration,
– resistance to strong shear forces, e.g. during pumping, and
– good dispersibility through the foaming medium, e.g. in coatings.

Silicones work in the same way but with the opposite effect in expanded plastic, especially PUR; i.e. they act as stabilizers. They distribute themselves uniformly throughout the foam, but, instead of displacing other foam-stabilizing surfactants, they assume the role of surfactants.

Unlike silicone defoamers, silicone foam stabilizers must be compatible with or even soluble in the foaming medium and capable of acting as a surfactant. Foam stabilizers are therefore made from modified siloxanes.

9.5. Silicone fluid emulsions

Silicone fluids are widely used in the form of aqueous emulsions. Emulsions allow easy dilution with water and homogeneous distribution of small quantities over substrates (hence their use as release and impregnation agents).

Emulsions of the oil-in-water type are preferred for silicone fluids.

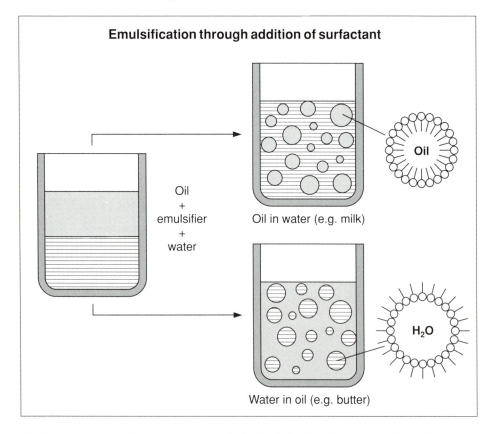

Stable dispersion of oil drops in water is effected by lining the oil-water interface with emulsifiers. A great many emulsifiers can be chosen from, all of which contain, as the example of dodecyl sulfate below shows, a lipophilic and a hydrophilic part.

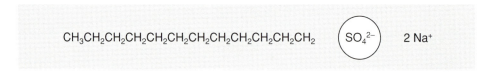

The residual charge on emulsifiers governs whether they are classified as anionic (e.g. sulfates), cationic (quaternary ammonium compounds) or non-ionic (ethoxylates). Covering the droplet surface with a layer of emulsifier reduces the surface tension and prevents the individual droplets from agglomerating. Another option is to stabilize the emulsion with stabilizers or thickeners.

Emulsions are prepared from silicone fluids of medium or low viscosity.

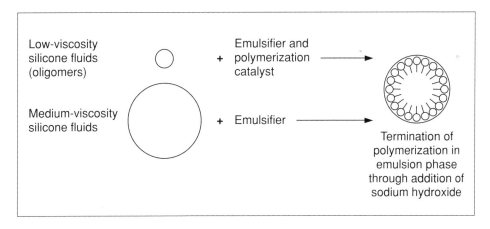

Emulsions differ considerably in particle size. Indeed particle size governs the stability of emulsions to a large extent.

Type of emulsion	Particle size
Fine	ca. 250 µm
Coarse	400 µm
Antifoaming	4 000 – 10 000 µm

Another means of characterizing aqueous emulsions of silicone fluids is afforded by values for the hydrophilic-lipophilic balance (HLB).

Substance	Use in emulsions w/o	Use in emulsions o/w	Use as solubilizer
Ether, oils			12 – 18
Waxes		10 – 12	
Paraffin oils	8	11	
Silicone fluids		11 – 12	
Wash-active substances			13 – 15

9.6. Release agents

There are various types of silicone release agents, e.g. oils, pastes, rubber coatings, release papers. Most prevalent of all, however, are the liquid silicone release agents, whose outstanding advantages are ready film formation – even in aqueous solution – and extreme anti-adhesive properties. Release agents made from silicone fluids show no tendency to accumulate in moulds, are effective in very low concentrations and do not impair the appearance of the finished article. However, they have to compete against a large number of less expensive release agents (such as waxes), metallic salts, polymers (PVA and PE), and powders (talcum and mica).

One unassailable advantage over organic release agents, however, is that they have a much more pronounced release action pro rata in application.

10. Silicone rubber

Silicone rubbers consist essentially of silicone polymers and fillers. They are mostly differentiated according to the type of polymer employed and the crosslinking mechanism.

Properties	Solid rubber (HTV* rubber)	Liquid rubber	RTV** rubber
Polymer (viscosity)	Pasty (solid) 20 million mPa s	5 000 – 100 000 mPa s	200 – 10 000 mPa s
Molecular weight	400 000 – 1 million	10 000 – 100 000	20 000
Chain length (silicone units)	10 000	1 000	200
Crosslinking mechanism	Mainly peroxide-induced at elevated temperatures	Mainly addition at elevated temperatures	Condensation and addition at room temperature

Fillers are either reinforcing or non-reinforcing.

Reinforcing fillers are primarily fumed silicas with a surface area in excess of 125 m^2/g. They owe their reinforcing action to the hydrogen bonds formed between the silanol groups at the surface of the silica (3 – 4.5 SiOH groups per mm^2) and the oxygen atom of the silanol groups on the α, ω-dihydroxy polydimethyl siloxane chain. This filler-polymer interaction increases viscosity and changes the glass-transition and crystallization temperatures. The polymer-filler bonds also enhance mechanical properties but may be responsible for premature crepe hardening of the rubber.

*HTV: high-temperature-vulcanizing, **RTV: room-temperature-vulcanizing

Non-reinforcing fillers interact extremely weakly with the silicone polymer. Examples include chalk, quartz flour, diatomaceous earth, mica, kaolin (china clay), $Al(OH)_3$ and Fe_2O_3. The particles have a diameter of around 0.1 µm. Their function is to raise the viscosity of the unvulcanized compounds and to increase the Shore hardness and modulus of elasticity of vulcanizates. They can additionally be surface-treated to augment tear strength.

Talcum occupies a middle position between reinforcing and non-reinforcing fillers.

Fillers are also utilized to produce special effects. For example, iron oxide, zirconium oxide and barium zirconate enhance thermal stability.

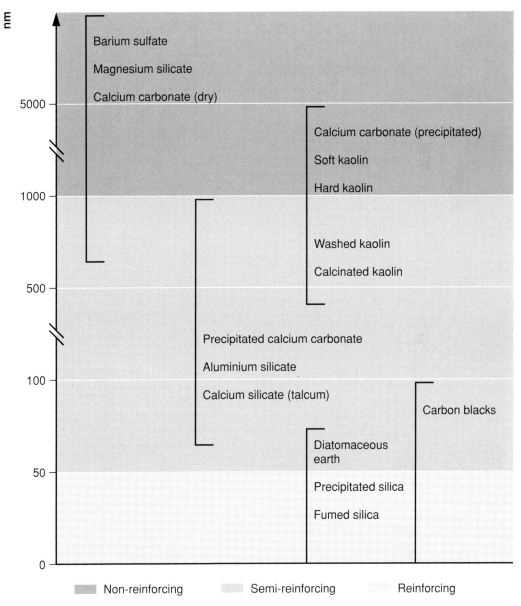

The influence of reinforcing and non-reinforcing fillers on selected mechanical properties is illustrated below.

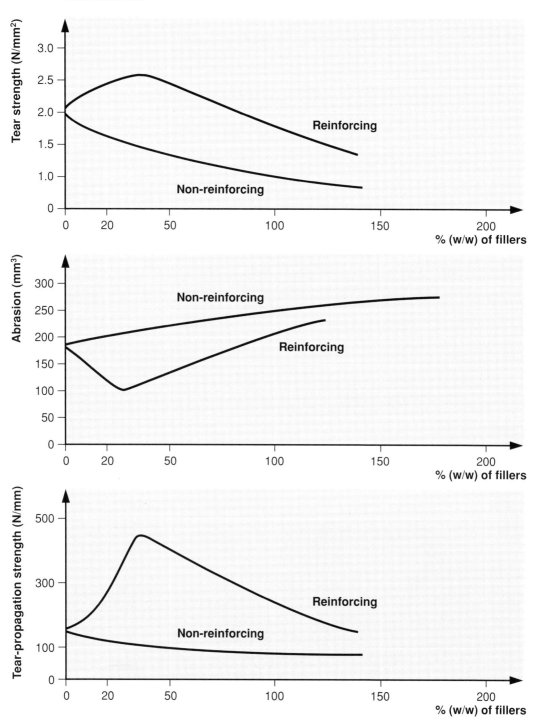

Further constituents of silicone rubbers can be catalysts, crosslinking agents, pigments, anti-adhesive agents, plasticizers and adhesion promoters.

Plasticizers in particular are needed for adjusting a low modulus of elasticity. Internal adhesion promoters consist of functional silanes that interact with the substrate on the one hand and the crosslinking silicone polymer on the other. They are mostly used in conjunction with RTV-1* rubbers.

Premature crepe hardening is pre-empted by low-molecular or monomeric silanol-rich compounds, such as diphenyl silanediol and water. They prevent the silicone polymers from interacting too strongly with the silanol groups of the fillers by reacting more rapidly with the fillers. A similar effect can also be achieved by partially covering the filler with trimethylsilyl groups (treating the filler with methylsilanes).
The siloxane polymer can furthermore be modified chemically to yield phenyl polymers or boron-containing polymers or be blended with organic polymers, such as styrene-butadiene copolymers.

Crosslinking of silicone rubber in the rubber industry is normally effected at room temperature or higher.

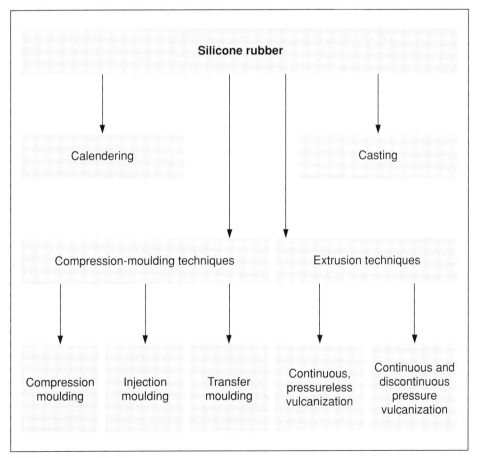

*RTV-1: a one-component RTV rubber

Crosslinking density is a key factor affecting the properties of the vulcanizate. It is greatly dependent on the crosslinking mechanism. Its influence on mechanical properties is shown below.

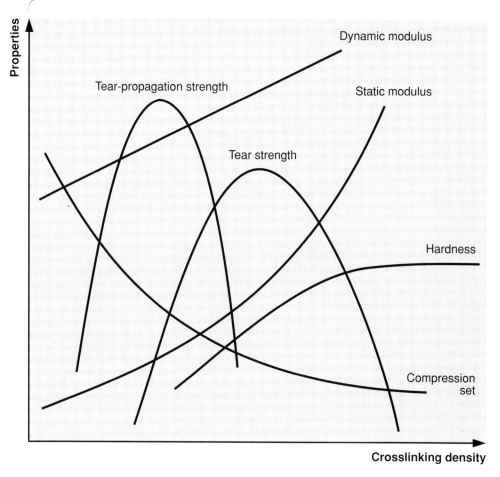

RTV silicone rubber crosslinks by both condensation and addition mechanisms.

Condensation crosslinking occurs between α,ω-dihydroxy polydimethyl siloxane and silicates in the presence of dibutyl tin dilaurate or tin (II) octoate catalysts. The rate of crosslinking depends on the functionality and concentration of the crosslinker, its chemical structure and the nature of the catalyst. The reaction is greatly accelerated by water.

Addition crosslinking involves addition of SiH across double bonds in the presence of catalytic salts and complexes of platinum, although palladium and rhodium are also suitable. The reaction proceeds at room temperature when platinum metal/olefin complexes are used. Addition-crosslinking rubbers for high-temperature processing contain nitrogenous platinum complexes, e.g. of pyridine, benzonitrile, or benzotriazole.

There are several major differences between these two types of curing mechanisms that have a bearing on the uses to which the rubbers are put (see diagram below).

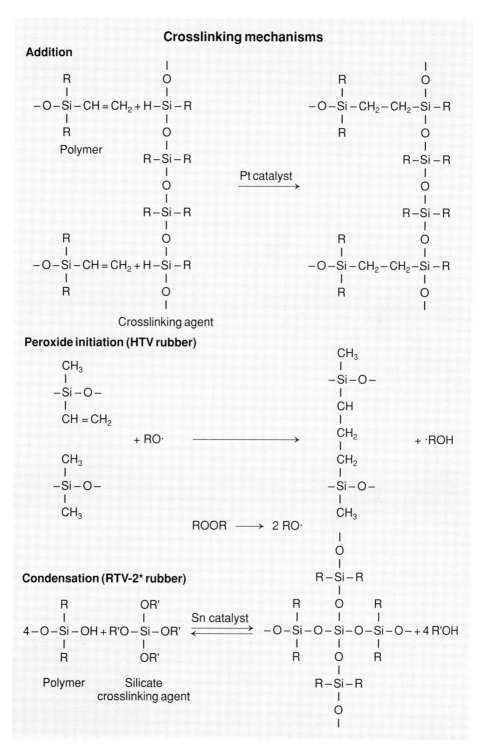

Differences between condensation and addition crosslinking RTV-2 rubbers

Condensation crosslinking	Addition crosslinking
Mixing ratio of silicone rubber and accelerator is variable within limits	Mixing ratio of both components is fixed
Crosslinking agent and catalyst contained in accelerator	Crosslinking agent (H-siloxane) in component 1; Catalyst (platinum complex) in component 2
Vulcanization impaired only by lack of water	Vulcanization impaired by various substances (sulfur compounds etc.)
Temperature has little effect on rate of vulcanization	Temperature greatly affects rate of vulcanization
Chemical shrinkage through elimination of alcohols during crosslinking	Virtually no shrinkage
Reversion (depolymerization) through elimination products (alcohols) possible above 80°C	Reversion impossible
Long processing time causes long vulcanization times	Rapid curing possible at elevated temperatures when processing time is long

RTV-1 rubber crosslinks by methods very similar to that of RTV-2. Systems that vulcanize only on use, e.g. when pressed from a tube, can be formulated at room temperature in a vacuum by choosing less reactive agents.

Crosslinking at elevated temperatures is effected by addition (for liquid rubber), peroxide initiation (for solid rubber) or, a recent development, highly energetic radiation. Sulfur is not used to crosslink silicone rubber because the vulcanizates have poor mechanical properties.

Peroxide crosslinking relies on the formation of free radicals to initiate the reaction. The incorporation of 0.05 – 1.0 mole % vinyl groups into the polymer produces more selective crosslinking and hence better vulcanizates.

Radiation crosslinking also involves free radicals, with doses of 2 – 5 Mrad being employed.

10.1. Properties of silicone rubber

Silicone rubber has several special properties that require closer inspection.

10.1.1. Elastic behaviour

The elastic properties of a material are best assessed by means of a stress-strain diagram.

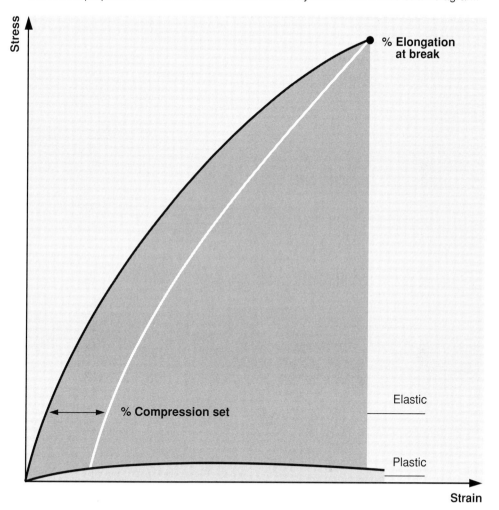

An important parameter used to describe elastic behaviour is the E modulus (modulus of elasticity). It is given by the stress-strain diagram as resistance to deformation in N/mm^2, expressed in terms of a specific strain value in %.

In particular, the elastic properties of silicone rubber change only very slightly as a function of temperature. In this respect, it differs from other elastomers.

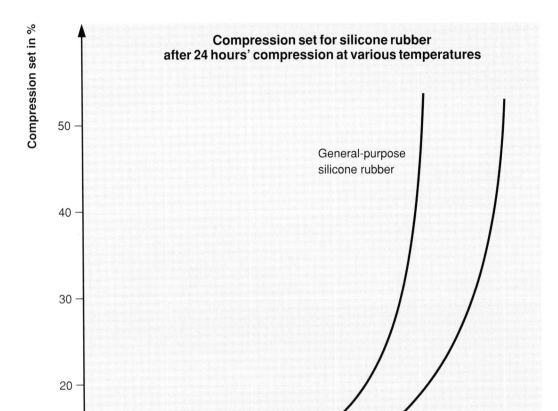

As shown above, the elastic properties, namely modulus of elasticity and compression set, are governed to a large extent by the crosslinking density. A further important contribution to elasticity is made by the composition of the blend, above all by the type and quantity of filler.

In the case of HTV silicone rubber, there is an additional dependence on the nature and proportion of crosslinking agent. For instance, the modulus of elasticity and the rebound resilience increase in line with the amount of crosslinking agent (see page 46).

It should be noted that temperature has very little effect on the compression set of silicone rubber.

10.1.2. Resistance to heat and cold

Silicone rubber is more resistant to the effects of temperature, especially to cold, than are a great many other elastomers. It is very stable towards ozone and radiation, a property which manifests itself in excellent resistance to weathering.

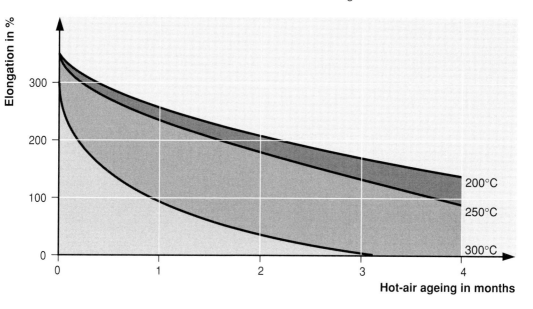

The resistance of silicone rubber to elevated temperatures is shown below.

At 150°C	2 – 4 years
At 200°C	1 year
At 250°C	100 days
At 300°C	about 14 days

Temperatures of up to 180°C cause only very gradual changes in its properties. Certain additives, such as carbon black, iron (II) oxide, and metal acids, imbue the rubber with exceptional resistance to temperatures of up to 250°C. If exposure is very brief, even temperatures as high as 900°C can be withstood.

The foregoing information pertains to heating in air. In closed systems, i.e. absence of air, the rubber reverts within a relatively short time, becoming progressively softer in the process.

Silicone rubber can withstand steam of up to 130°C for long periods of time. Consequently, medical devices containing silicone rubber can be sterilized in steam without any problems arising. A high crosslinking density is reflected in enhanced resistance to steam. The effect of temperature on tear strength is shown on page 53.

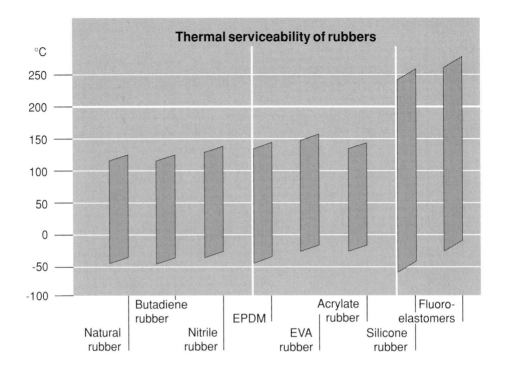

Irreversible impairment of mechanical properties does not occur at temperatures below 200°C. For example, when the tear strength drops temporarily by around 50 % on exposure to a temperature of 200°C, the original value can still be regained, provided that the silicone polymer has not undergone thermal degradation.

Particularly noteworthy is the low compression set at elevated temperatures.

Silicone rubber retains its elasticity even at extremely low temperatures. Special low-temperature blends can withstand exposure to -60°C to -120°C. The phenyl groups contained in the blends greatly depress the crystallization and glass-transition temperatures.

Silicone rubber is highly resistant to ozone. For example, 70 hours' exposure at 40°C to an atmospheric ozone concentration of 2 % does not impair any of its properties. It is also remarkably resistant to γ-radiation and X-rays. If 50 % ultimate elongation is taken as a criterion for measurement, silicone rubber can withstand 40 – 50 Mrad; and special phenyl types, up to 100 Mrad.

The corollary of this high resistance to ozone and radiation is exceptional weathering properties. Even after 5 years' outdoor exposure, silicone rubber will only have suffered a 50 % drop in tear strength.

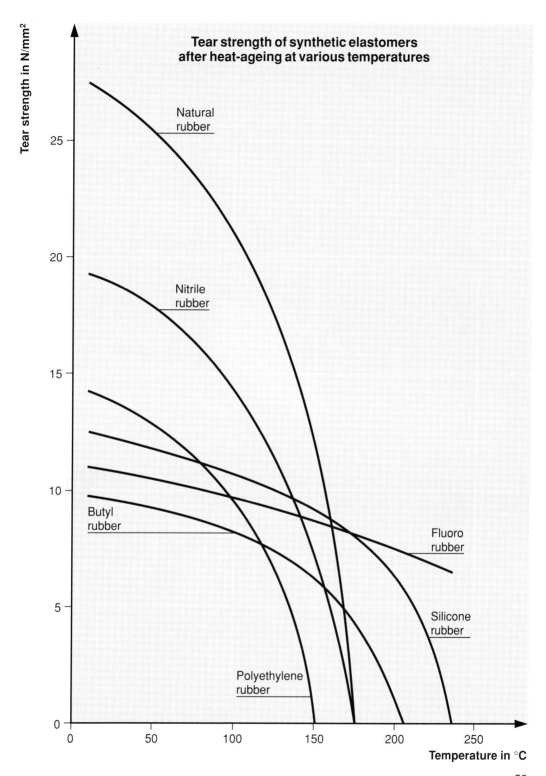

10.1.3. Mechanical properties

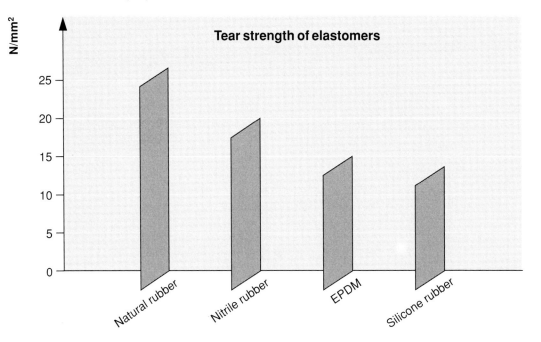

As can be seen from the diagram, specially formulated silicone rubber has tear strengths of over 10 N/mm². This improvement on the values for the standard types (tear strengths: 5 – 6 N/mm²) is achieved by replacing methyl by vinyl polymers and using fumed silica. It should be noted that there is a simultaneous improvement in tear-propagation strength.

Any comparisons with organic elastomers should take account of mechanical properties at elevated and low temperatures. In this respect, those of silicone rubber are found to change only very slightly with temperature. Indeed, silicone rubber surpasses organic elastomers in mechanical strength at temperatures greater than 200°C.

Abrasion resistance and the rate at which cracks form merit special attention. Values of 10 – 35 N/mm for tear-propagation strength ($\approx$ 10 N/mm² tear strength) correspond to abrasion-resistance values* of 200 – 1600 rev/cm (after immersion in oil) and crack-propagation rates** of 120 000–150 000 cycles/cm. In evaluating crack propagation, allowance must be made for possible superposition of tear and tear-propagation strengths.

*ASTM D 1630/61, **ASTM D 813/59

10.1.4. Flame retardancy

Silicone rubber has excellent flame retardancy. It has a flash point of 750°C, an ignition temperature of 450°C and its LOI* values lie well below the critical value of 21 % that approximates the concentration of oxygen in air.

Only very minor amounts of smoke are evolved during combustion. More importantly, combustion does not release any toxic or aggressive gases, such as HCl and sulfur compounds.

$$\begin{array}{c} CH_3 \\ | \\ -O-Si-O \\ | \end{array} + O_2 \longrightarrow \begin{array}{c} CH_2OOH \\ | \\ -O-Si-O \\ | \end{array} \longrightarrow SiO_2 + CH_2O$$

$$-CH_2OOH + O_2 \longrightarrow CO_2 + H_2O$$

The principal combustion products are carbon dioxide and water. The SiO_2 burns to an ash that is an excellent dielectric and renders silicone rubber one of the most reliable insulating materials. For this reason, siliconized cables are used in such critical sectors as shipbuilding, aircraft construction and public buildings.

The combustion behaviour of several elastomers is tabulated below.

Polymer	LOI %	Smoke density	Fire load (MJ/kg)
Polychloroprene (CR)	36.5	65	20.9
Chlorosulfate polyethylene (CSP)	30.7	85	15.9
Nitrile/PVC blend (NBR/PVC)	30.6	136	20.5
Silicone rubber (VMQ)	26 – 42	45	15.9
Styrene/butadiene rubber (SBR)	21.6	265	17.6
Ethylene/propylene terpolymer (EPR)	23.5	280	36
Chemically crosslinked polyethylene (XLPE)	20	205	41.9
Poly(vinyl chloride) (PVC)	24.3	265	22.2

Particularly flame retardant grades of silicone rubber are made by addition of $Al(OH)_3$ hydrate or minute amounts of platinum compounds and TiO_2 (for self-extinguishing products).

*Limiting oxygen index

10.1.5. Electrical properties of silicone rubber

Silicone polymers have outstanding dielectric properties and consequently silicone rubber is one of the best insulators available. Added to which, the electrical properties vary only slightly as a function of temperature.

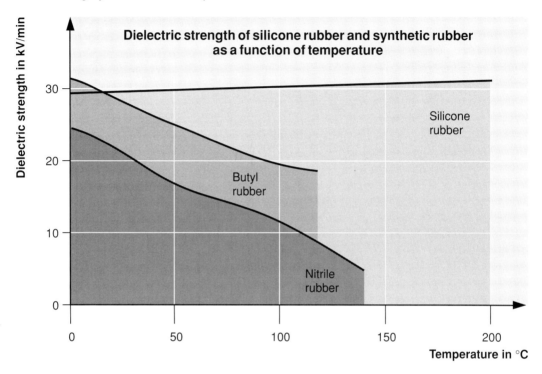

One outstanding electrical property of silicone rubber is its tracking resistance, which lies in the range 2.5 – 3.5 kV/6h. Silicone rubber has an arc resistance of approximately 250 s and a corona resistance of about 40 kV. Qualities that are particularly resistant to high voltages can be obtained by increasing the crosslinking density, i.e. raising the content of the crosslinking agent.

In contrast, electrically conducting blends can be prepared by adding conductive fillers, such as carbon black. This extends the range of conductivity enormously.

The electrical conductivity of such rubbers varies with applied pressure, a property which is exploited in pressure sensors.

Resistivity of selected substances

Resistivity in Ω cm	Substance	Silicone rubber
10^{18}	PTFE, PE	Insulating silicone rubber, cable insulation, electrical insulation
10^{15}	PP	
10^{12}	Polyamide, PVC	
10^{9}	Cellulose (paper, cotton)	
10^{3}	Rubber blends with carbon black, metal fibres	
1	Carbon fibres, conductive silicone rubber	150 Ω cm (field homogenization) 2 Ω cm (contact mats) 0.01 Ω cm (electromagnetic field control)
10^{-3}	Doped polyacetylene	
10^{-6}	Copper, silver	

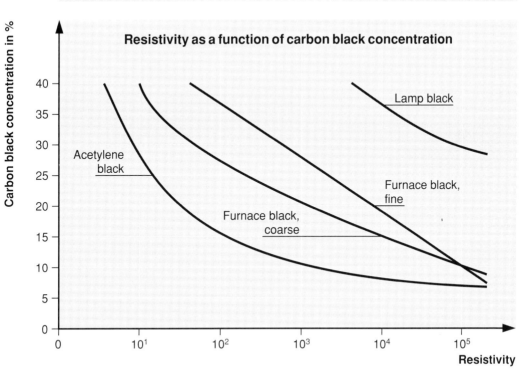

Resistivity as a function of carbon black concentration

10.1.6. Resistance to oil and chemicals

Silicone rubber is generally resistant to chemicals and can be used in applications involving contact with dilute acids and alkalis. However, its chemical resistance decreases with increase in concentration or temperature.

Its resistance to various oils compares favourably with that of organic elastomers. The following changes in volume can be expected after 14 days' immersion in hot oil (150°C).

Resistance of silicone rubber to solvents, fuels and oils

Chemicals	% Increase in volume
Solvents and fuels (after 7 days at RT)	
Acetone	15–25
Carbon tetrachloride	>150
Ethyl alcohol	0–10
Iso-octane	>150
Xylene	>150
Oils (after 14 days at 150°C)	
ASTM Oil 1	5
ASTM Oil 2	8
ASTM Oil 3	40
SAE Oil 20 W 20	25
Silicone fluid, viscosity 100 mPa s	30

Aggressive oil components, both the oxidizing and decomposed types, do the most damage to silicone rubber. Apolar solvents, such as white spirit and benzene, and chlorinated hydrocarbons cause it to swell extensively. However, the original properties are restored when the solvent has evaporated. Silicone fluids also cause reversible swelling of silicone rubber, with the least swelling effected by the phenyl types. It should be noted that silicone rubbers containing trifluoropropyl groups are very resistant to oil.

10.1.7. Release properties / Adhesion to substrates

The surface of silicone rubber is extremely anti-adhesive (i.e. non-stick), a fact which is reflected in its having a much lower surface energy than have organic polymers.

Polymer type	Surface energy (J)
Polydimethyl siloxane	21 – 22
Polyphenylmethyl siloxane	26
Poly(vinyl chloride)	40
Polyethylene	30
Starch	40
Wool	45

The low surface energy is retained even when the rubber has a high filler content, e.g. of hydrophilic silica.

Silicone rubbers differ markedly in their adhesion to substrates.

– RTV-1 rubbers are generally very adhesive.

– RTV-2 rubbers do not adhere to most materials.

– HTV rubbers possess intermediate adhesive properties.

In some cases, adhesion can only be accomplished by pre-treating the surface with a primer. For instance, the primers required for metals and plastics are based on resins and silane adhesion promoters. Alternatively, adhesive groups can be incorporated into the rubber components themselves.

The anti-adhesive and perfectly smooth surface of silicone rubber as well as its non-toxic properties lie behind its use in the medical field. For instance, infusion and transfusion tubes are made from silicones because they prevent blood from coagulating.

10.1.8. Gas permeability

The low intermolecular forces present in silicone polymers confer mobility on the polymer chains, even on those that are crosslinked. One consequence of this is the high gas permeability of silicone rubber relative to other elastomers.

Elastomer	Gas permeability $\dfrac{10^6 \text{ cm}^3}{\text{s cm}}$
Silicone rubber	60
Natural rubber	2.4
Butyl rubber	0.24
Poly(vinyl chloride)	0.014
Teflon	0.0004

The high gas permeability of silicone rubber is utilized in coatings that can breathe and to separate gases.

10.1.9. Flow properties of silicone rubber

Because it flows so well, silicone rubber is easily processed, e.g. by casting and extrusion techniques. Its thixotropic properties can be very readily adjusted with fumed silica.

Extrusion techniques call for particularly good rheological properties, e.g. in cable production. The flow properties of silicone rubbers can be described by the energy of activation.

Polymer type	Activation energy kJ/mol
Methyl polysiloxane	14.2
Phenylmethyl polysiloxane	50.0

Care must be taken to avoid crepe hardening, which can seriously affect flow and is caused by hydrogen bonding between the polymer and the OH groups of the fillers.

Crepe-hardened rubber mixtures can be liquefied by mill-freshening, which ruptures the hydrogen bonds.

Modern silicone rubber blends of the non-milling type are virtually free from crepe-hardening. They are made by either adding short-chained OH-polymers that primarily orient themselves towards the surface of the filler or using fillers that have been treated with silanes.

10.1.10. Damping properties

Low modulus of elasticity combined with a high mechanical dissipation factor renders silicone rubber an excellent medium for absorbing shock (impact) and sound. Its rebound resilience remains almost constant from room temperature up to 200°C.

Damping coefficients of silicone polymers can exceed 90 %.

As can be seen from the following diagram, silicones can be compressed more than organic polymers.

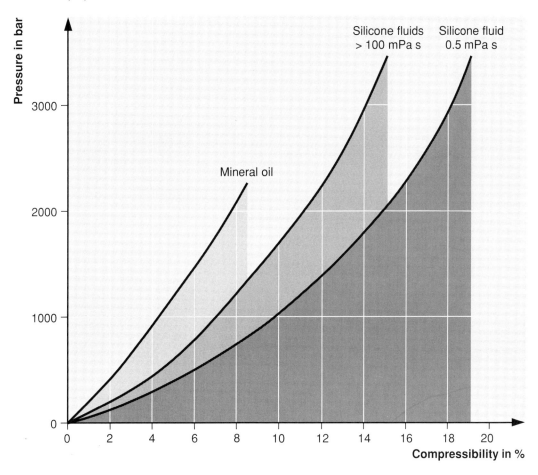

More importantly, the permanent change in volume after protracted periods of cyclic load is extremely slight (compression set).

When shock or impact occurs, mechanical energy is converted to thermal energy and heat begins to build up. However, since silicone rubber is thermostable, this has very little effect on it and the rubber therefore has a very long fatigue resistance.

Severe shock can rupture the crosslinks in silicone rubber, converting some of the material to fluid. This effect is reversible, however, and crosslinking slowly occurs again afterwards.

Certain boron-containing polymers readily crosslink via boron-oxygen linkages. Shock immediately converts them from this slightly crosslinked state to the fluid state.

10.1.11. Versatility through compounding

Selected properties of silicone rubber can be enhanced by judicious compounding. The following table shows the mechanical properties of the most important types of silicone rubber.

Properties Rubber type	Hardness Shore A	Tear strength N/mm^2	Tear-propagation strength N/mm	Compression set %
Solid rubber Multipurpose grades	40–70	8	17	35
High mechanical strength	40–70 40–70	10 10	18 35*	30 30
Low compression set	40–70	7	15	10
No-post-cure grades	40–70	7	15	25
Grades for cable and wire insulation	60–70	8	25	
Oil-resistant grades	70	5	25	35
Liquid rubber	40	10	30	15**
RTV rubber (addition and condensation curing)	30	5	20	< 25

The best values for tear strength are afforded by a low content of vinyl groups and the use of treated, i.e. silanized, fillers.

---

* Although two rubbers may have the same tear strength, they may differ in tear-propagation strength.
** Very high values for tear-propagation strength and compression set in liquid rubber.

10.2. Processing of RTV silicone rubber

RTV-2 silicone rubber starts to vulcanize as soon as the two components are blended. The viscosity increases progressively until a rubber-elastic state is reached. The period of time during which the rubber remains pourable is called the pot life.

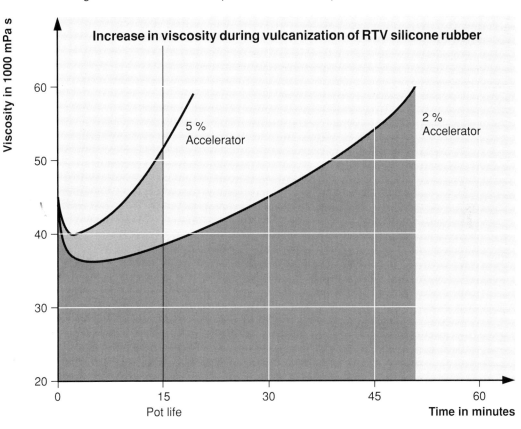

The rate at which condensation-curing rubbers vulcanize depends on the quantity and the type of accelerator, which itself contains silicates and a catalyst (curing rate increases from methyltriethoxy silane to tetraethyl silicate). The curing rate of condensation types is greatly accelerated by traces of added water whereas that of addition types varies according to the catalyst, which is usually a platinum compound.

Some shrinkage must be expected when RTV silicone rubber is being processed. Condensation types shrink due to the elimination of reaction products (alcohol). Addition-curing rubbers hardly shrink at all.

A distinction is drawn between shrinkage that occurs during vulcanization, due to elimination of alcohol, and that which occurs in usage.

Shrinkage behaviour is illustrated in the following diagram. Total shrinkage is of the order of 0.5 – 2 % for condensation types and considerably less than 0.1 % for addition types.

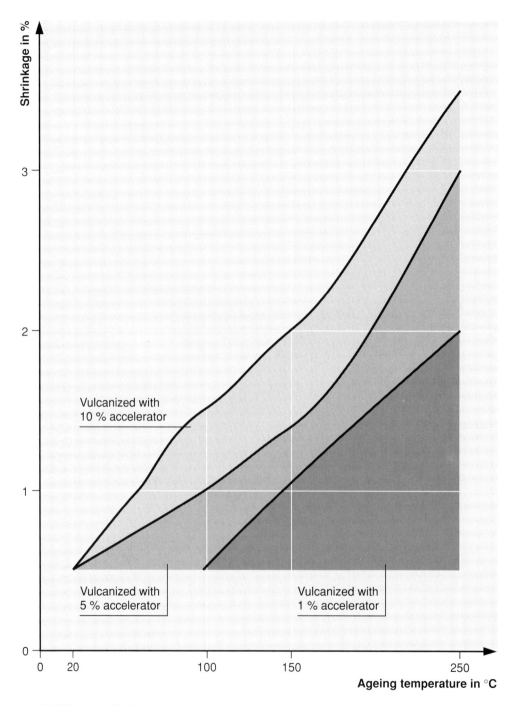

Addition-crosslinking can be halted at the gel stage to yield a tacky, weakly crosslinked vulcanizate that has good inherent adhesion. The gelatinous state arises from the fact that chain growth between the H- and vinyl groups of the bifunctional polymers is competing with crosslinking.

When most of the vinyl polymer is replaced by OH polymer, addition-crosslinking results in foamed silicone rubber. As the following reaction scheme shows, it is the hydrogen evolved during the reaction that causes the rubber to foam.

$$\begin{array}{c} | \\ O \\ | \\ O-Si-OH \\ \diagup \end{array} + \begin{array}{c} | \\ H-Si-O- \\ | \\ O \\ | \end{array} \longrightarrow \begin{array}{c} | \\ O \\ | \\ O-Si-O-Si-O- \\ \diagup \quad\quad | \\ \quad\quad O \\ \quad\quad | \end{array} + H_2\uparrow$$

The foam has primarily a closed-pore structure.

RTV-1 rubbers cure by means of condensation, which is initiated by atmospheric moisture, e.g. when the product is pressed from a tube. The rate at which they cure depends to a large extent on the humidity of the air.

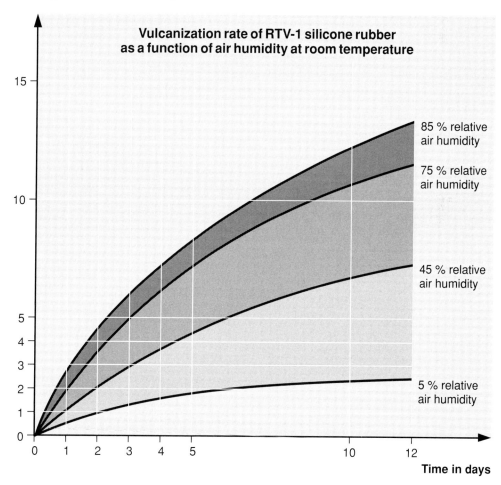

Vulcanization rate of RTV-1 silicone rubber as a function of air humidity at room temperature

The various uses of RTV silicone rubber are shown below.

RTV-1 rubber	Bonding	Building joints
	Sealing and encapsulating	Sealing of industrial articles (automobiles, mechanical engineering)
RTV-2 rubber	Sealing and encapsulating	Engine seals, electrical insulation, insulators, cable accessories
	Coating	Electrically insulating tapes, printing techniques (Tampoprint)
	Mould-making	Ceramic and souvenir articles, electronic components
	Release agents	PU foams, baking tins and trays

10.3. Processing of HTV silicone rubber

Heat-curing silicone rubber can be vulcanized both batch-wise in autoclaves and continuously by extrusion. The continuous method is particularly important and can be carried out either under pressure or at atmospheric pressure in hot air. Continuous, pressurized techniques include injection and transfer moulding, and steam vulcanization in a vulcanization tube.

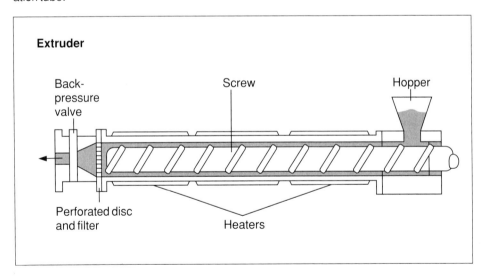

Other techniques include vulcanization in a fluidized bed, in a salt bath and radiation-curing. Cable insulation is extruded by a special technique in which the silicone rubber is applied to the cable core via an angular extruder head.

The greatest single factor governing each of these techniques is the crosslinking agent. Peroxide initiators, which are the primary agents for hot curing, differ in the temperature required to decompose them and thus initiate vulcanization. The most important peroxides are tabulated below.

Crosslinking agent	Application	Amount of crosslinking agent	Decomposition products pH range	Efflorescence
Bis(2,4-dichloro-benzoyl) peroxide 50 % in silicone fluid	Extrusion (pressure-less vulcanization at 250–450°C)	1.1–1.8 %	Acidic	White powder
	Compression vulcanization 100–150°C	1.2–1.5 %		
Bis(benzoyl) peroxide 50 % in silicone fluid	Compression vulcanization Steam vulcanization 110–150°C	1–1.5 %	Acidic	White crystals
t-Butyl perbenzoate	Compression vulcanization 140–160°C	0.5–1 %	Acidic	None
Dicumyl peroxide (95 %)	Compression vulcanization 150–170°C Steam vulcanization	0.6–0.9 %	Neutral	None
2,5-Bis(t-butyl-peroxy) 2,5-dimethyl-hexane (95 %)	Compression vulcanization 160–180°C	0.25–0.5 %	Neutral	None

The following points are of particular importance in connection with peroxides.
– The temperature at which the peroxide decomposes governs the storage stability of the rubber blends (risk of crepe hardening at low decomposition temperatures).
– The elimination products generated can affect the properties of the vulcanizate. For example, acidic elimination products formed during compression vulcanization may lead to reversion.
– The use of dicumyl peroxides for initiating hot-air vulcanization leads to tacky vulcanizates.

The extent of vulcanization can be followed from rheometer curves. Starting from a certain initial Mooney viscosity, the viscosity drops briefly and then increases for a period lasting between 10 seconds for addition crosslinking and 10 minutes for peroxide crosslinking, ending ultimately in a rubber-elastic state. In the final stages, both over-curing and reversion of the silicone rubber can occur (see diagram on page 68).
Vulcanization generally requires post-curing for 2 hours in hot air at 200°C. The purpose of post-curing is to
– remove elimination products from the vulcanizate, and
– reduce compression set.

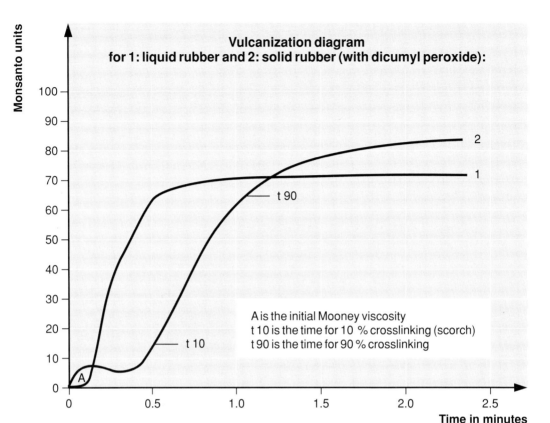

The elimination products generated during vulcanization are the main causes of reversion. Acidic elimination products can be neutralized with selected alkaline reversion stabilizers.

A comparison of the two most important extrusion techniques is shown below.

Hot-air vulcanization	Steam vulcanization
Low investment costs for equipment	High production rate (200 m/min)
Vulcanization occurs at once (low risk of crepe hardening)	Vulcanizates have good compression set and high-voltage resistance

Hot-air vulcanization is primarily adopted for silicone rubber intended for tubes, profiles and cables whereas steam vulcanization is resorted to only for inexpensive, i.e. highly filled, cables.

Flexible foams made from HTV rubber are obtained by addition of special pore formers. The foams are mainly open-pored if the blowing agent decomposes below the vulcanization temperature and closed-pored if it decomposes above.

The following table classifies the various kinds of silicone rubber by chemical substituent.

Basic types of silicone rubber

Polymer type	Symbol	Temperature range °C	Comments
Dimethyl	MQ	-70 to 250	High compression set, low thermal stability under permanent load, high peroxide concentration during vulcanization
Vinylmethyl	VMQ	-70 to 250	Low compression set, greater thermal resistance under permanent load, better vulcanization behaviour
Phenylvinyl-methyl	PVMQ	-100 to 250	Good low-temperature resistance, better flame retardancy, lower oil resistance
Fluorovinyl-methyl (fluorosilicone)	FVMQ	-60 to 180	Very great resistance to oils and aromatic solvents, low compression set

There are five basic kinds of silicone rubber – characterized by the type of rubber, the composition of the blend and the vulcanization process – and each one is suited to a specific application.

Type	Preferred vulcanization	Application
1. Transparent blends (contain reinforced fillers, have high tear strength)	Compression Extrusion	Moulded articles and tubing for medical applications, the food sector and high-tech products
2. Blends for a broad range	Compression	Seals
3. Oil-resistant blends (contain non-reinforcing fillers, have low compression set)	Compression	Seals
4. Cable blends (good electrical properties)	Extrusion	Cable and wire insulation
5. Electrically conducting blends	Extrusion and compression	Ignition cables, contact mats

In addition, the processing conditions can often be extensively manipulated to modify the rubber's properties, e.g. to yield steam-resistant or high-voltage-resistant vulcanizates.

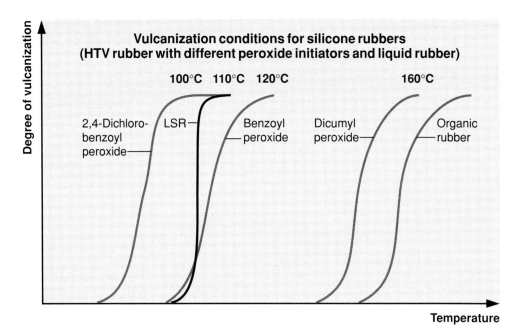

These vulcanization curves clearly illustrate the extreme rapidity with which liquid silicone rubber (LSR) vulcanizes. A comparison of solid and liquid rubber reveals the following points.

Solid rubber	Liquid rubber
Extruded (rubber industry)	Injection moulded LIM technique (plastics industry)
Inexpensive products	
Numerous compounding recipes to suit special requirements	Inexpensive processing (no preparation of blends, shortest cycle times)

The way in which HTV silicone rubber is processed governs its fields of application.

Solid rubber	Extrusion Calendering	Cables, profiles, tubes, fabric coating
	Compression moulding Injection moulding	Seals, membranes
Liquid rubber	Injection moulding	

11. Silicone resins

Resinous properties are displayed by silicones that mainly contain T and Q units.

11.1. Properties and types of silicone resins

Silicone resins are usually high molecular substances (molecular weights: 2 000 – 4 000) and are thus primarily supplied as solutions in organic solvents. However, solvent-free liquid and silicone resin powders are also available.

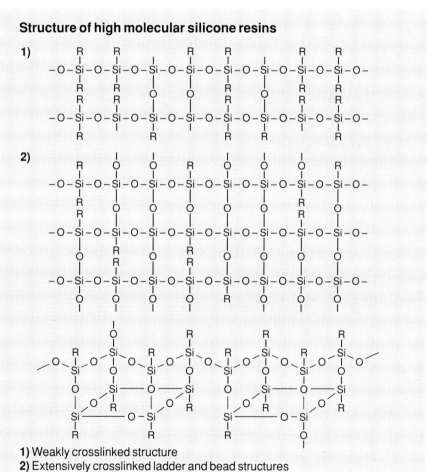

1) Weakly crosslinked structure
2) Extensively crosslinked ladder and bead structures

As can be seen from the diagram above, silicone resins are networks of primarily trifunctional units arranged irregularly. A high proportion of bifunctional units confers greater elasticity and a rubber-like structure. Regular arrangements are found in so-called polysesquioxanes, which have a quartz-like structure.

Similar resin structures can be obtained by using low-molecular trifunctional siloxanes (molecular weights: 500 – 2000). These types of resin intermediates have the structure shown below.

```
                    R              R
                    |              |
                   Si ——— O ——— Si
                  / |            | \
                 O  OH          HO  O
                /                    \
         R – Si – OH            HO – Si – R
                \                    /
                 O  OH          HO  O
                  \ |            | /
                   Si ——— O ——— Si
                    |              |
                    R              R
```

Only silicates are purely quadrifunctional. They revert to pure silica on curing.
Silicone adhesive resins also contain these quadrifunctional units.

11.2. Processing of silicone resins

When the solvent evaporates from a silicone resin solution, a dry resin film is left behind. However, this physical drying process hardly involves any polycondensation at all, with the result that subsequent heating yields a tacky, soft mass. To effect complete curing, the resins are subjected to temperatures above 200°C for several hours.

The rate of curing can be accelerated somewhat through the use of catalysts, such as zinc octoate, lead and cobalt naphthenate, and butyl titanate.

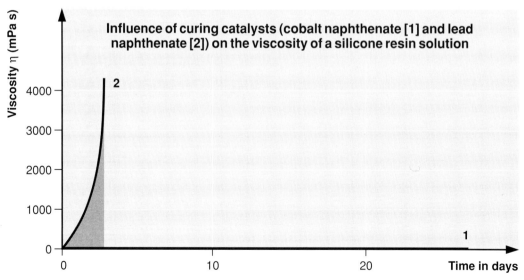

Influence of curing catalysts (cobalt naphthenate [1] and lead naphthenate [2]) on the viscosity of a silicone resin solution

When these catalysts are used, however there is a risk that the catalysed resin solutions will gel prematurely before use. Adding alcohols as stabilizers alleviates this problem. The alkoxy groups generated are less liable to condense.

$$\gtrless SiOH + ROH \longrightarrow \gtrless Si-OR$$

Apart from condensation-curing resins, there are peroxide-curing resins and addition-curing resins. The latter type is particularly important for solvent-free, low-viscosity encapsulants.

[Reaction scheme: Oligomer with Si–H and Ph, CH$_3$ substituents reacts with Oligomer bearing CH=CH$_2$ groups; with Heat and Catalyst yields $-Si-CH_2-CH_2-Si-$ linkage]

The following survey shows the most important areas of application for **trifunctional siloxanes** (silicone resins).

Alkoxy siloxanes (silicone intermediates)	Binders for paints Heat-resistant binders	Varnishes
	Crosslinkers	Silicone rubber
Silicone resins	Binders for paints Heat-resistant binders	Varnishes Electrical insulation
	Impregnating agents	Masonry protection
	Encapsulants	Encapsulation of electric motors
	Release agents	Plastics industry Electrical industry Baking tins / trays

12. Chemicals, petrochemicals and coal industries

Silicones pervade the entire chemicals industry, whether as raw materials, pharmaceuticals, man-made fibres or consumer products.

Let us now turn our attention to some of the major areas in which silicones play a crucial role.

12.1. Recovery and refining of crude oil

The use of silicone defoamers starts as early as the extraction of crude oil from bore holes. This is particularly true of the so-called tertiary process in which strongly foaming surfactants have to be added to the drilling fluids.
But severe foaming also occurs at subsequent stages in the refining process, as described in the following sections.

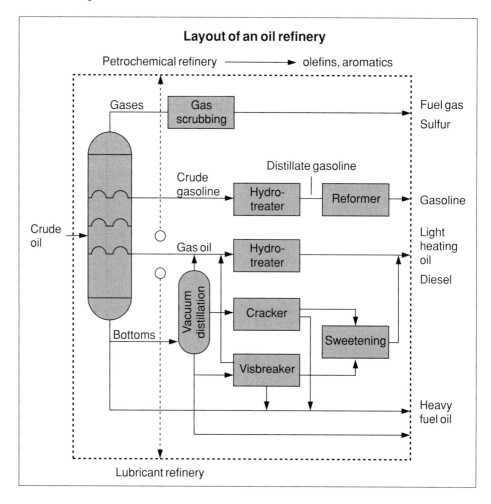

Layout of an oil refinery

12.1.1. Gas scrubbing

Scrubbing of gases with aqueous solutions of ethanolamine can cause foaming owing to reaction with dissolved H_2S.

12.1.2. Separation of aromatics

The various techniques employed to separate aromatic hydrocarbons, such as benzene and toluene, from crude oil fractions, crack gases and hydroformates are all more or less prone to foaming.

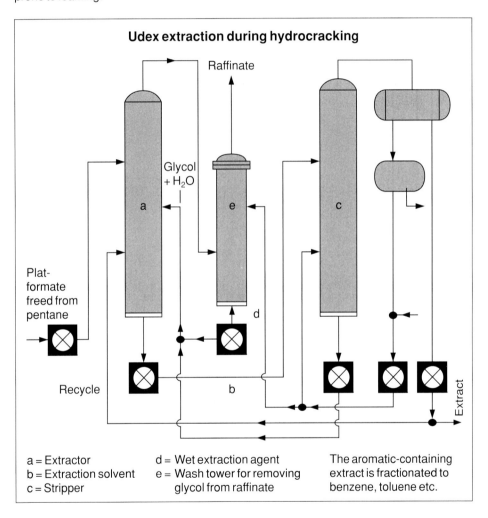

Udex extraction during hydrocracking

a = Extractor
b = Extraction solvent
c = Stripper
d = Wet extraction agent
e = Wash tower for removing glycol from raffinate

The aromatic-containing extract is fractionated to benzene, toluene etc.

In the Udex extraction process, the water-diethylene glycol mixture used as solvent has a strong tendency to foam.

The Shell "Sulfolane" and Snam Progetti "Formex" processes employ sulfolane and N-formyl morpholine respectively as solvents.

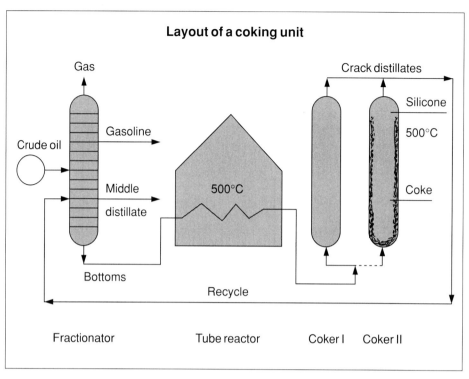

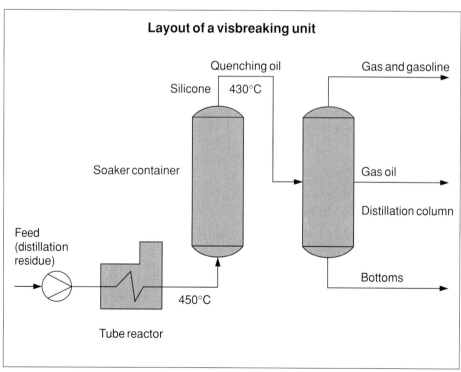

12.1.3. Cracking

The purpose of cracking is to break up hydrocarbons into molecules of lower molecular weight.

$$CH_3-CH_2-CH_2-CH_2-CH_2-CH_2-CH_3 \longrightarrow CH_3-CH_2-CH_2-CH_3-+CH_2=CH-CH_3$$

The catalysts required for fluidized catalytic cracking consist of zeolites (alumino-silicates) dispersed in a matrix of clay and silica gel, the latter being derived from ethyl silicate.

Other cracking processes, such as visbreaking, employ milder conditions and mainly yield light heating oil and gasoline. Cracking by the coking and delayed-coking techniques involves higher temperatures and longer dwell times, resulting in complete conversion of distillate residues in cokers and calciners to coke and volatile components. Steam cracking is used to manufacture basic chemicals, such as ethylene, for the petrochemicals industry.

12.1.4. Refinery processes

Cracking is followed by refining processes and hydrogenation (hydrotreating), which also offer scope for using defoamers.

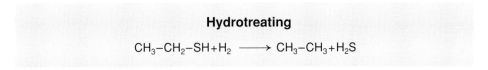

Hydrotreating

$$CH_3-CH_2-SH+H_2 \longrightarrow CH_3-CH_3+H_2S$$

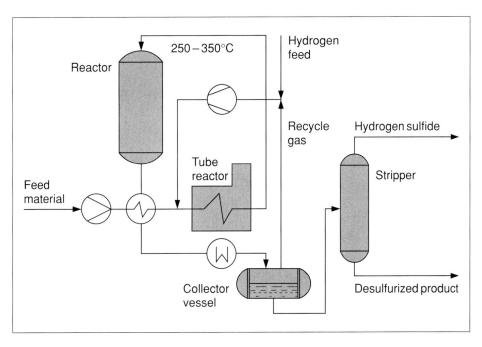

In hydrotreating, organosulfur compounds are converted to hydrocarbons and hydrogen sulfide. At the catalytic reforming stage, the hydrocarbons are subjected to brief cracking in order to dehydrogenate the naphthene rings*. This allows the octane value of fuels to be raised.

Silicones improve the quality of some fuels and propellants. For example, the efficiency of ship engines is enhanced by adding silicones to the fuel oils. Coal briquettes are coated with silicone resin dispersions to prevent dusting.

12.2. Chemical feedstocks

Foaming occurs in many manufacturing processes for chemical feedstocks. Agrochemicals are a prime example.

During the manufacture of phosphoric acid as a raw material for fertilizers (see equation below), $CaCO_3$ present as an impurity in the phosphates gives rise to foaming.

$$Ca_3(PO_4)_2 \xrightarrow{HNO_3 + H_2SO_4} H_3PO_4 + Ca(NO_3)_2 \text{ or } CaSO_4$$

The reaction products are used to manufacture a fertilizer consisting of $Ca(H_2PO_4)_2$ and $Ca(NO_3)_2$ (Nitrophoska®).

Another crucial reaction in the production of fertilizers is impeded by foam.

$$NH_4NO_3 + Ca(OH)_2 \longrightarrow CaNH_4NO_3 \text{ (at 180°C)}$$

Additional foaming often occurs when the excess ammonia is stripped off.

Defoamers are frequently added to agrochemicals to prevent foaming from occurring during application. For instance, a wide variety of herbicides and insecticides are prone to this problem, whether applied as dispersions or as wettable powders ("flowables").

Various silicone defoamers are added to them to counteract the foaming tendency. A further application of defoamers arises in the production of sodium hydroxide by the diaphragm process. The liquor is evaporated at 350°C to leave behind a thin layer of solid NaOH.

Explosives are embedded in viscous substances, such as silicone rubber, in order to desensitize them.

Silanes are also used to synthesize pesticides in the agrochemicals industry. Silicone surfactants have proved to be excellent wetting and spreading agents, and silicas are used as thixotropic thickening agents and flow improvers.

* Term used in the petroleum industry for cycloalkanes

12.3. Polymerization

Applications in polymer chemistry include that of defoaming and the use of silanes as initiators, crosslinkers and catalyst modifiers.

12.3.1. Defoamers for polymerization processes

In polymerization processes, such as suspension polymerization of PVC, the presence of surface-active substances, such as poly(vinyl alcohol), at the expansion stage, gives rise to foaming of unreacted vinyl chloride. If defoamers are not used, there is a loss of yield and more work is required for cleaning the equipment. Hydrophilic silicas are often added to the finished granules as antiblocking agents and flow improvers.

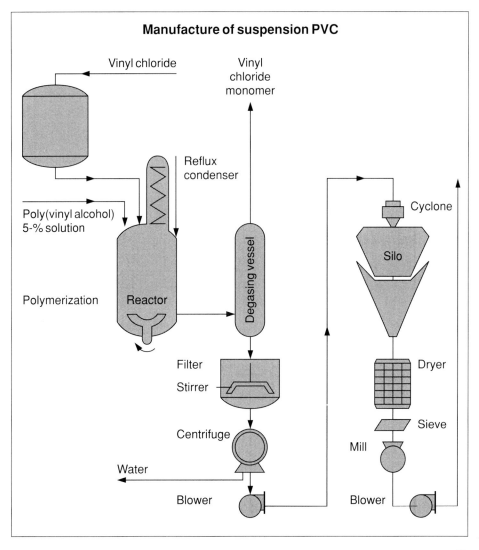

12.3.2. Polymerization with silanes

Being reactive substances, silanes can participate in various ways in polymerization processes and thus be incorporated into organic polymers. Foremost here is their use as cross-linking agents (cf. post-cured polyethylene in the plastics industry).

Alkoxy silanes are used in the polymerization of ethylene to introduce siloxane units, resulting in better control over product properties.

Certain organosilicon compounds initiate polymerization of acrylates and methacrylates. In group-transfer polymerization, the reaction propagates by transferring the organosilicon groups from monomer to monomer.

Silanes can also act as stabilizers for many organic polymers.

Hydrogen-containing silanes are commonly used as reductants. In one polymerization technique for propylene, H-siloxane is employed instead of a Ziegler-Natta catalyst.

12.3.3. Silane-modified catalysts

The use of silanes to modify Ziegler-Natta catalysts for manufacturing polypropylene has been a major step forward. The improved catalysts are more reactive and result in a greater yield of isotactic polymer.

$$H_2C=CH-CH_3 \xrightarrow{\text{Catalyst}}$$

Atactic:
$$-CH_2-\underset{\underset{CH_3}{|}}{CH}-CH_2-\underset{\underset{CH_3}{|}}{CH}-CH_2-\underset{\underset{CH_3}{|}}{CH}-CH_2-\underset{\underset{CH_3}{|}}{CH}-$$

Syndiotactic:
$$-CH_2-\underset{\underset{CH_3}{|}}{CH}-CH_2-\underset{\underset{CH_3}{|}}{CH}-CH_2-\underset{\underset{CH_3}{|}}{CH}-CH_2-\underset{\underset{CH_3}{|}}{CH}-$$

Isotactic:
$$-CH_2-\underset{\underset{CH_3}{|}}{CH}-CH_2-\underset{\underset{CH_3}{|}}{CH}-CH_2-\underset{\underset{CH_3}{|}}{CH}-CH_2-\underset{\underset{CH_3}{|}}{CH}-$$

The improvements were made in several stages with the result that 4 generations were developed for different processes, e.g. gas-phase and bulk polymerization.

- The first generation comprised the classic Ziegler-Natta catalysts, such as $TiCl_3 + Al(C_2H_5)_3$. Yield of isotactic polymer: 80 – 90 %.

- In the second generation, "internal electron donors", such as ether, raised the isotactic yield to 96 – 98 %. However, the catalysts' reactivity was disappointing.

- In the third generation, $MgCl_2$ was employed as a solid carrier on which aromatic carboxylic acids acted as electron donors.

- By the fourth generation, the $MgCl_2$ was prepared by precipitation and not by milling. At the same time, silanes acted as "external electron donors". These catalysts have the advantage of being low-dusting and universally applicable, e.g. in the Spheripol and Hypol processes.

The silanes used for the fourth generation catalysts are alkoxy silanes, such as phenyltriethoxy silane, diphenyldimethoxy silane and cyclohexylmethyldimethoxy silane.

12.4. Pharmaceutical products

The introduction of the silyl group $Si(CH_3)_3$ into organic compounds, a process known as silylation, opened up a whole new field of organic syntheses.

Silyl groups can be incorporated permanently into organic compounds (e.g. to yield "sila pharmaceuticals") or can serve as temporary protective groups during syntheses. Silylation has assumed great importance in the production of antibiotics.

12.4.1. Silylation reactions

Temporary incorporation of the silyl group changes the properties of intermediates and thus improves the synthesis route. Particular changes are as follows.

- Increased volatility: thus making it possible to analyse compounds of low volatility, such as oligosaccharides and steroids, by gas chromatography.

- Enhanced solubility in organic solvents: trimethylsilylated cellulose is used to decrease the polarity of groups such as $-OH$, $-NH_2$.

- Temporary protection of sensitive functional groups, such as hydroxyl and carboxyl: the silyl group is cleaved simply by hydrolysis.
  A few examples are given below to illustrate the great extent to which this protective function is exploited in syntheses.

The most important silylation reaction is that employed in the synthesis of ß-lactam antibiotics. Natural precursors are first used to make synthetic derivatives of penicillin and cephalosporin.

The synthesis of ampicillin from 6-APA is shown below.

Silylation also affords a means of converting penam to cephem derivatives.

*N,N'-bis(trimethylsilyl)urea

The way in which antibiotics function explains the demand for further derivatives.

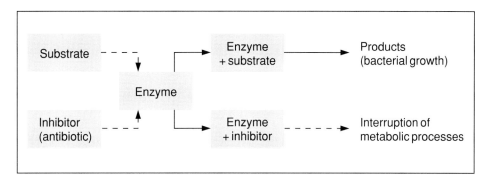

Antibiotics function by inhibiting enzymes and thereby interrupting bacterial growth. However, the emergence of new, more resistant strains has spurred on development work and led to a prolific number of antibiotics. These are termed fourth generation. The greater number of steps involved (as many as 15) in synthesizing these antibiotics has promoted the use of new silylating agents. For example, whereas trimethyl chlorosilane is the most common silylating agent for ampicillin, tertiary butyl compounds are resorted to for third-generation antibiotics, such as cefotaxim, and fourth-generation types such as inipenem and aztreonam.

Several important silylating agents are listed below.

Silylating agent	Application area
$(CH_3)_2SiCl_2$ Dimethyl dichlorosilane	Inexpensive silane. Disadvantages: e.g. difficult phase separation of reaction products (the dihydroxy siloxane product forms emulsions).
$(CH_3)_3SiCl$ Trimethyl chlorosilane	Broad-spectrum silylating agent. Yields highly defined silylation products, hydrolyses to readily separable products.
$(CH_3)_3SiNHSi(CH_3)_3$ Hexamethyl disilazane	Silylation proceeds without formation of by-products (only ammonia evolved).
$H_3C-C(=O)-NH-Si(CH_3)_3$ N-Trimethylsilyl acetamide	Allows silylations at temperatures lower than those with hexamethyl disilazane. Highly selective.
$CO[NHSi(CH_3)_3]_2$ Bis-(trimethylsilyl)urea	As above, but features greater silylating power.
$(H_3C)_3C-Si(CH_3)_2Cl$ t-Butyldimethyl chlorosilane	Forms silyl ethers that are particularly stable towards hydrolysis.

Hydrolysis of the trimethylsilyl group yields hexamethyl disiloxane, whose Si–O–Si linkage can be cleaved by HCl to regenerate $(CH_3)_3SiCl$ (cf. Recombination).

A further application of silylating agents is the synthesis of prostaglandins for regulating blood pressure.

Silylation of testosterone and progesterone prolongs their hormonal activity.

Many pharmacological syntheses have only become feasible with the advent of silyl catalysts. One example is the preparation of nucleosides via silylated N-bases in the presence of a catalyst such as trimethylsilyl perchlorate.

R = o-benzyl
Ac = acetyl

Nucleosides are important cytostatic agents.

Siliciferous catalysts can also be derived from triphenyl silanol, such as triphenylsilyl vanadate. This catalyst is used in the so-called Rupe rearrangement for the synthesis of citral, a raw material for perfumes.

Acetylene derivative $\xrightarrow{(C_6H_5)_3SiOH \ + \ VOCl_3}$ Aldehyde (citral)

12.4.2. Sila pharmaceuticals

Substitution of silicon atoms for carbon atoms has led to what are known as sila pharmaceuticals of which much is expected in many areas. Unlike traditional drugs, they afford the possibility of limiting the efficacy of drugs to a defined period of time since the Si–O–C bond is susceptible to hydrolysis and constitutes a weak point in the molecule. Such drugs are thus readily degradable.

Sila pharmaceuticals have been developed that show antispasmodic, analgesic effects, are effective against certain types of carcinomas, and inhibit formation of cholesterol.

12.5. Man-made fibres

Silicones improve the performance characteristics of man-made fibres and are used as auxiliaries at several stages in their production, whether by wet or dry spinning techniques.

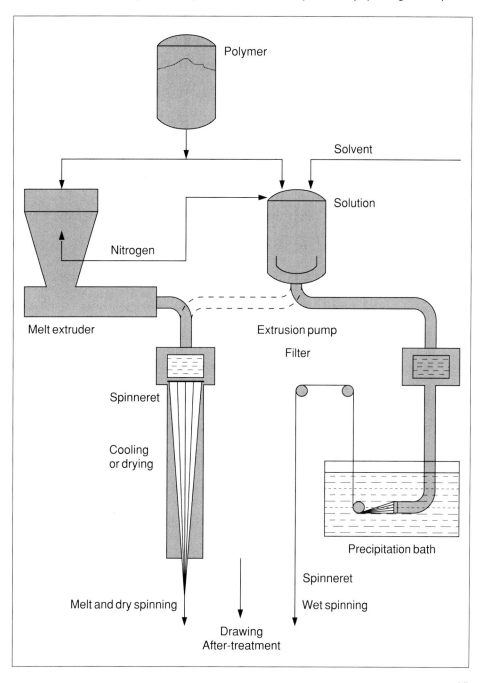

The following diagrams illustrate the production lines for acrylic and polyester fibres, which are made by polyaddition and polycondensation techniques respectively.

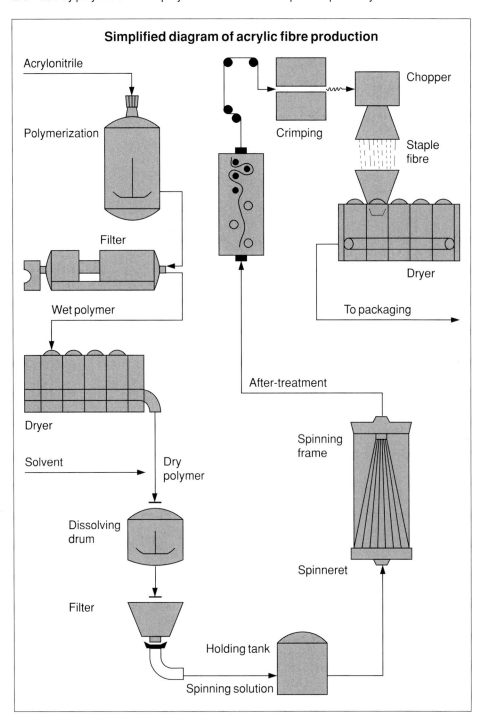

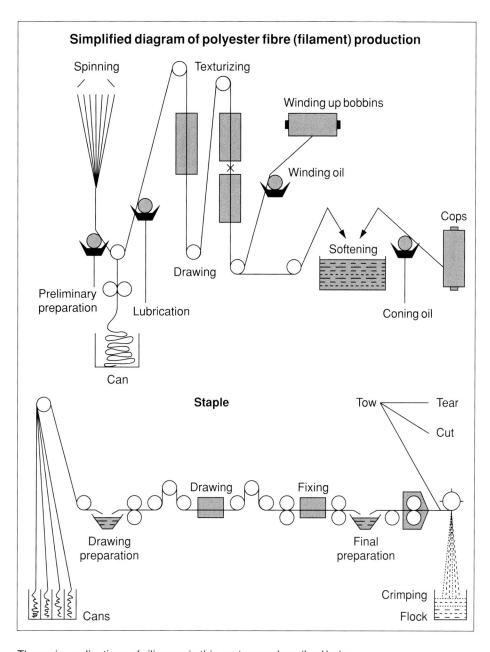

Simplified diagram of polyester fibre (filament) production

The main applications of silicones in this sector are described below.

They are added during polycondensation processes to confer special effects on the polymer fibres, such as internal antistatic properties, antipilling properties, better drawability and permanent smoothness.

A heat-resistant lubricant is required at the exit orifice of the fibres from the spinneret for which application silicone fluid sprays have proved ideal.

Foaming in spinning baths, such as a viscose precipitation bath, necessitates the use of antifoams. Special silicone fluids are added to prevent the fibres from sticking during reactive spinning of polyurethane fibres.

Very important roles are played by silicone fluids in lubricants, especially those fluids which are employed during texturizing and drawing processes.

The demands imposed on them are manifold. They must lubricate and withstand temperatures in the range 200 – 250°C. Furthermore, they serve as spreading agents for other ingredients and must prevent ingredients that have decomposed at lower temperatures from adhering to the fibres. Once they have fulfilled these tasks, they must wash out readily from the preparations so as not to impair subsequent dyeing processes. For this reason, water-soluble, glycol-modified silicone fluids are employed. In many instances, these also act as release agents when staple fibres are being cut.

Man-made fibres are modified in order to imbue them with similar wearability as natural fibres. Polyester fibres have good tear strength and crease resistance and are used most in the clothing sector. Polyamide fibres have good tear and abrasion resistance and find use as nylon stockings and home textiles. Acrylic fibres can be rendered as bulky as wool and are used for pullovers etc.

A high degree of bulk can be achieved by permanent finishes based on crosslinkable silicone emulsions. Siliconized staple fibres are suitable fillers for cushions and linings for overcoats and anoraks. Treatment with the silicones usually takes place after crimping in the fibre-production process.

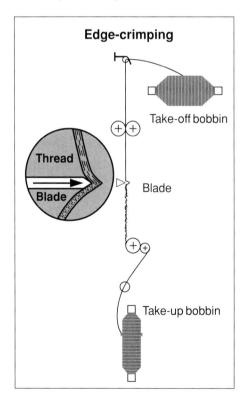

The texturizing process is worth a special mention because the fibre surface is deliberately roughened by friction.

Texturizing can be performed in a number of ways (torsion, stuffer box, air-lay system, edge-crimping).

There is a tendency nowadays to amalgamate spinning, stretching and texturizing and this has led to processes such as stretch / spinning texturizing and high-speed spinning.

Increases in production rates from 100 m/min for texturizing to as much as 4000 m/min for high-speed spinning clearly demonstrate the need for highly efficient spin finishes.

12.6. Treatment of pigments and fillers

The enormous variety of particle sizes and compositions of pigments and fillers requires a wide range of chemicals to treat them, e.g. paraffins, silicones, silanes.

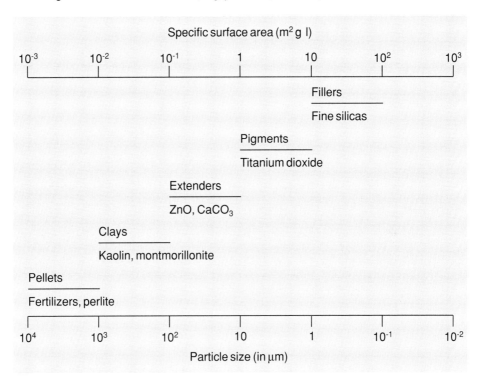

The effects that can be achieved depend, particularly in the case of silanes, on the absorption capacity of the powder concerned. They follow the order below.

None	Carbon black, graphite
Weak	Zinc, lead, gypsum, lime
Good	Asbestos, iron, talc, mica, inorganic oxides
Excellent	Silicas, quartz, glass, organic silicates, aluminium, copper

Siliconization has two different effects:

- enhanced flow and dispersion, e.g. in treatment with silicone fluids and methylsilanes, and
- improved adhesion between fillers and polymers, e.g. in treatment with functional silane adhesion promoters.

The first category embraces treatment of pigments (TiO_2, Cr_2O_3) with silicone emulsions for thermoplastics processing. Treating hygroscopic powders with silicones improves flowability. The most important applications in this case are $NaHCO_3$ powder fire extinguishants that have been treated with H-siloxane. The flow properties of powders can be improved by adding anticaking agents (hydrophobic silica). Extensive use of this is made in fertilizers, blasting powders etc.

The second category includes fillers for reinforcing plastics and rubbers. Here, intimate contact between polymer and filler is crucial.

The rubber industry mainly employs fillers, such as silicas, kaolin, quartz flour and aluminium oxide trihydrates that have been treated with silanes bearing pendant SH –, NH_2– or vinyl groups.

In the plastics industry, silanized glass fibres are the most common reinforcing agents. Glass fibres are used extensively in industry, finding application in rubber tyres, thermally insulating plasters, PVC floor coverings and electrical insulation tapes. The glass fibres are silanized as they are being made, e.g. by mechanical drawing, blast drawing and crystal pulling. The manufacturing process resembles that for man-made fibres.

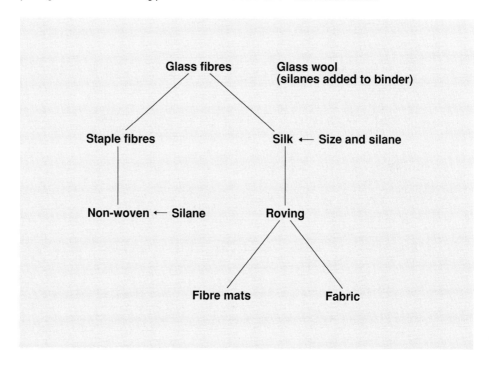

The Sol-Gel process can be used to make special fillers from alkoxy silanes. This yields a route to products that have a much broader range of properties than glass.

Such organically modified silicates, with their alternating silicate and organic structural units, are comparable to filler-reinforced plastics. They have much better thermal, mechanical and optical properties and are used in the electronics industry for circuit boards and in the plastics industry for scratch-resistant coatings. If functional silanes are treated by the Sol-Gel process, the surfaces of the resultant fillers are covered to a very specific extent with functional groups.

Polycondensation of functional silanes affords a way of preparing heterogeneous catalysts of metal complexes on a silicate matrix. Catalysts of this kind are used to effect hydrosilylation, carboxylation and hydrogenation.

The Sol-Gel process also yields ion exchangers with very defined spherical surfaces.

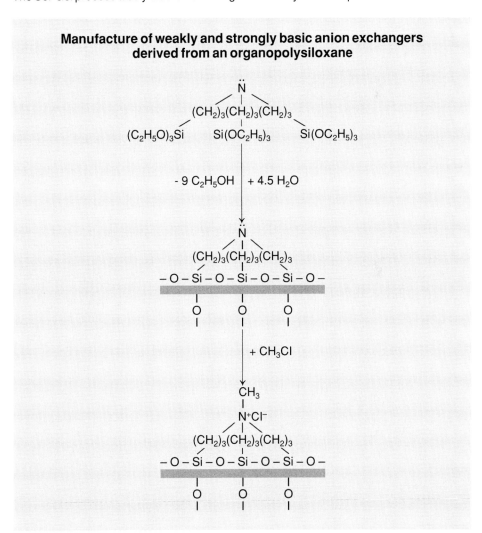

Manufacture of weakly and strongly basic anion exchangers derived from an organopolysiloxane

12.7. Chemical consumer products

The chemicals industry offers a large number of consumer products whose properties have been extensively improved through the addition of silicones.

12.7.1. Washing powder

The need to protect the environment has led to changes in the composition of washing powders and detergents, with the main aim being to find a substitute for phosphates.

Composition of washing powders	
Wash-active substances ("Syndets")	e.g. fatty alcohol sulfates ethylene oxide adducts
Phosphates	partially replaced by alternative products, e.g. zeolites, citrate, nitrilotriacetate
Optical brighteners and bleaches Wetting agents	

Owing to severe eutrophication, i.e. over-enrichment of lakes in organic and mineral nutrients, attempts are being made to substitute so-called builders for triphosphates. The function of these builders is to soften the water by forming complexes with the salts responsible for the hardness, to enhance the power of the surfactants to dissolve dirt and to prevent the washing powder from caking.

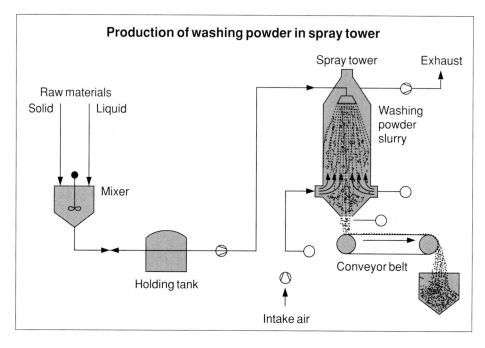

Production of washing powder in spray tower

Alternative substances are in many ways inferior to phosphates and so greater amounts of surfactants have to be added. As a result, more foam is generated in the washing machine and resort must be made to silicone defoamers.

The defoamers are best added to the washing powder slurry prior to the spray-drying stage. However, if they are too finely emulsified, their activity is impaired. Furthermore, they must not decompose in the alkaline slurry.

Soft rinses for washing machines often contain silicone softeners in order to improve the handle of the goods.

12.7.2. Cleaners and polishes

Car polishes benefit greatly from added silicone fluids. Methyl silicone fluids impart higher gloss and make polishing easier. They function as internal lubricants for the wax particles, which can thus be distributed much more easily over the substrate. Combinations with aminomethyl silicone fluids are the most common types because the amino groups adhere to metallic surfaces, thereby increasing the polishes' resistance to being washed off.

Silicone fluids are also added to a number of other types of polishes, including those for furniture, glass-ceramic cooker hobs, and plastic fittings in cars.

12.7.3. Cosmetics

Cosmetics profit from silicones as well.

The ability of silicone fluids to form films that spread easily is highly valued in skin-care cosmetics. Silicone fluids are extremely compatible with the skin and leave it feeling soft and silky. They promote dermatologically active substances and prevent skin creams from being washed off too quickly. The latter property is of particular interest for sunscreen agents.

Hair-care cosmetics, i.e. shampoos, that contain aminofunctional silicone fluids add body to the hair and render it more pleasant to the touch. They also make wet hair much easier to comb.

When used in deodorants and antiperspirants, silicone fluids leave the skin feeling dryer and smoother. They keep the nozzles of powder sprays clean and prevent their clogging. Since they do not soil or leave grease marks on textiles, silicone fluids can be used as substitutes for oleous vehicles.

Silicone fluids are added to toiletries to impart a pleasant feeling to the skin and to reduce moisture loss. Silicone soaps are milder on the skin than conventional ones.

Benefits derived from silicones in the field of make-up and decorative cosmetics include reduced chalking of lipsticks, better dispersibility of powder formulations and enhanced flow properties of nail varnishes.

Toothpastes represent the most important field of application for fumed silica. The silica has a thickening action and is responsible for the pseudoplastic behaviour of the toothpaste, which flows from the tube under pressure only.

13. Plastics industry

Silicones represent only about 0.2 % of all plastics. They are, however, widely used for modifying plastics and as processing auxiliaries in the plastics industry.

13.1. Modification of plastics

Many organic plastics profit enormously from being post-cured, a crosslinking process that greatly enhances electrical and mechanical properties (e.g. elongation at break) and resistance to ageing by oxidation and radiation.

A prime example is PE, which is mainly crosslinked with vinyl silanes.

1. Grafting reaction (peroxide-initiated)

$$\begin{array}{c} | \\ CH_2 \\ | \\ HC\cdot \\ | \\ CH_2 \\ | \end{array} + H_2C=CH-Si(OR)_3 \longrightarrow \begin{array}{c} | \\ CH_2 \\ | \\ HC-CH_2-\overset{\cdot}{C}H-Si(OR)_3 \\ | \\ CH_2 \\ | \end{array}$$

2. Crosslinking

$$\begin{array}{c} | \\ CH_2 \\ | \\ CH-CH_2-CH_2-Si \\ | \\ CH_2 \\ | \end{array} \begin{array}{c} OR \\ | \\ \boxed{OH \; + \; H}O \\ | \\ OR \end{array} \begin{array}{c} OR \\ | \\ -Si-CH_2-CH_2-CH- \\ | \\ OR \end{array} \begin{array}{c} | \\ CH_2 \\ | \\ \\ | \\ CH_2 \\ | \end{array}$$

The reactions can be performed in several ways.

The Sioplas E method features two distinct reaction stages. The first consists in preparing a masterbatch of catalyst and grafted polyethylene and the second in mixing and extruding it.

In the Monosil process, grafted polyethylene and the catalyst mixture are extruded together and then crosslinked in the presence of water.

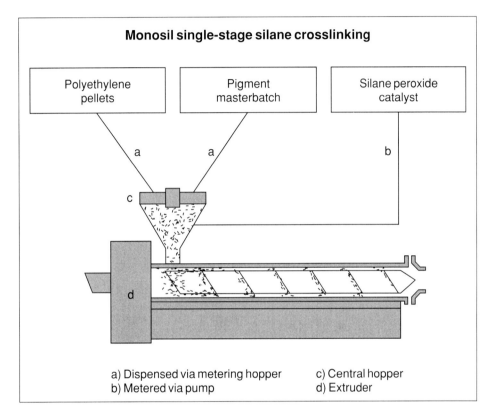

Monosil single-stage silane crosslinking

a) Dispensed via metering hopper
b) Metered via pump
c) Central hopper
d) Extruder

The thermal resistance of polyethylene is raised to such an extent by post-curing that XLPE* can compete against polypropylene as a material for the manufacture of hot-water pipes. The equally large, attendant improvement in its electrical properties has led to its becoming one of the most widespread dielectric materials.

PVC and polyamides are also post-cured, albeit to a lesser extent. The resultant plastics are used for tubes and cable insulation in the automobile industry.

Current research work is directed at incorporating siloxane groups into such thermoplastics as polystyrene and PVC with a view to modifying their properties.

Such modified thermoplastics have a lower coefficient of friction and better notched-impact resistance. Only a small proportion of silicone elastomer is needed to achieve these improvements.

* Crosslinked polyethylene

Silicone-modified thermoplastics are made by free-radical polymerization of monomers such as butyl acrylate, styrene, and vinyl acetate in a polydimethyl siloxane grafting matrix. If dihydroxy polysiloxanes are used, rubber-like products are obtained.

Interesting effects can be obtained by incorporating functional silanes. For instance, certain aminosilanes reduce electrical treeing in polyethylenes.

The performance properties of thermosets and thermoplastics are governed to a large extent by the fillers incorporated. Silanized glass fibres primarily serve to increase mechanical strength, e.g. impact resistance. They are added to unsaturated polyesters for skis and boat-building, to epoxy resins for compression-moulding compounds and printed circuit boards, and to phenolic resins for insulating materials. Silane (adhesion) bonds between the polymer and glass fibres provide the intimate contact required for obtaining the optimum properties.
But glass fibres are not the only fillers employed for reinforcing purposes. For instance, mica is used to reinforce polypropylene.

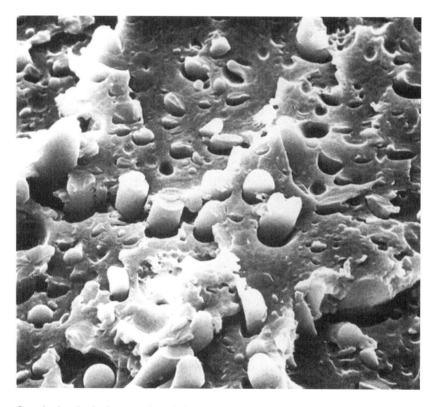

Certain rheological properties of plastics can be regulated by addition of fumed silica. The most important classes of products are plastisols for sealing the underbodies of automobiles, and unsaturated polyester resins. The latter are common surface-coating materials (e.g. laminating resins, gel coats, trowelling compounds), and fillers and must be thixotropic in many applications.

The scratch resistance of such plastics as polymethacrylates and polycarbonates can be enhanced by treatment with alkoxysilanes.

13.2. Auxiliaries for plastics processing

The chief purposes of silicones as plastics-processing auxiliaries are to stabilize foams and act as release agents.

Both rigid and flexible PU foams constitute the main application for silicone stabilizing fluids. Silicone stabilizers have also been developed for PVC foams, latex coatings and fire extinguishants. In total contrast, there are special silicone fluids which act as thermo-destabilizers in coagulation processes.

Silicone release agents are ubiquitous in the plastics industry. They are mostly based on silicone fluids, but silicone resins and rubbers are also used (in rotational moulding and for PU foam respectively).

They are processed by all types of compression and extrusion techniques and function as

– external release agents in moulds, and

– internal release agents (as additives to plastics pellets).

In the latter case, the silicone fluid, being incompatible, migrates to the surface of the plastic to act as a release agent.

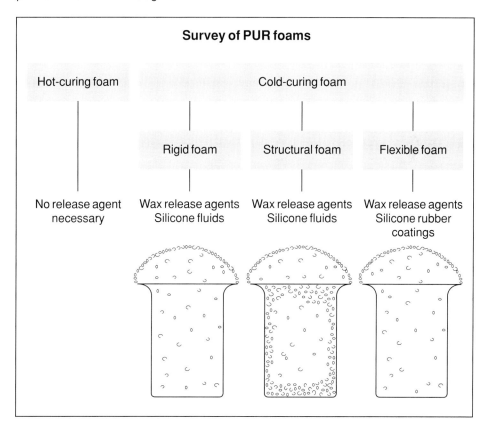

Release agents are crucial to expanded or foamed plastics, such as polyurethane. The most common uses of release agents are described below.

Structural foams, which have properties midway between those of rigid and flexible foams, are exceptionally strong due to their integral surface skin. They are processed by means of reaction injection moulding (RIM) in which foaming occurs in the mould itself.

$$-NC=O + H_2O \longrightarrow -\underset{\underset{H}{|}}{N}-\underset{\underset{O}{\|}}{C}-OH \longrightarrow -NH_2 + CO_2\uparrow$$

The release agents must neither affect the structure of the surface pores nor interfere with subsequent overcoating of the expanded articles.

Silicone-rubber mould-making compounds are widespread in the plastics industry. Moulds made from them can be used to make plastic replicas of nearly all objects (in epoxy resins, PU foams, imitation woods, polyester resins, imitation marble). Special techniques allow replicas to be made from moulds into which a skin of silicone rubber is inserted to act as the negative mould.

Fittings for automobiles are manufactured on a large scale by slush moulding. Production of artificial leather utilizes fabrics coated with silicone rubber and silicone release papers as carriers during the gelling process. Applying thermoplastics (PVC) to textured carriers is not the only way of making textured leather. Others include printing, varnishing and flock coating (applying particles to substrates coated with adhesive). One special technique for coating natural leather utilizes silicone rubber dies.

Silicone rubber dies are also used for high-frequency embossing.

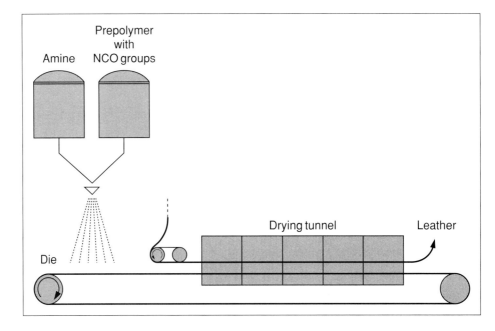

14. Rubber industry

Silicone rubber is the most commonly employed silicone product in the rubber industry, with a number of different silicone products also acting as auxiliaries in the production of organic elastomers.

14.1. Articles made of silicone rubber

Silicone-rubber articles perform a multitude of tasks in the most diverse industries. The most important articles are cable insulation (cf. Electrical industry), tubes (cf. Medicine), profiles and seals.

All kinds of silicone seals are called on to fulfil a myriad of tasks.

− In cars: rotary shaft seals, cylinder head gaskets, oil sump gaskets

− In domestic appliances: cookers, refrigerators and coffee percolators

− In the construction industry: windows

− In medical devices: syringes

The choice of sealing material depends not only on the operating conditions but also on the type of seal required, with HTV silicone rubber often competing against other types of silicone rubbers. Various types of seals are contrasted in the table below.

Flat seals	Liquid seals
1. Soft seals (rubber, plastic, paper, IT-seals) 2. Multi-material seals (rubber-metal fabric, metal with soft overlay)	RTV-1 silicone rubber
Easily installed and replaced Immediately fully operational	Lower material costs Smooth surface not important Seal highly reliable

A further consideration affecting the choice of material is whether the seal is of the static or dynamic type. The latter type, e.g. a rotary shaft seal, has to withstand extensive abrasion.

Other articles for which silicone rubber is the preferred material are caps and plugs for electrical fittings, diaphragms for breathing equipment, diver's goggles etc.

14.2. Silicone auxiliaries for the rubber industry

The main purpose of silicone auxiliaries is to surface-treat fillers and to act as release agents.

Silanization of fillers, such as fumed silica, enhances mechanical properties, including those of silicone rubber. Partial silanization of the surface prevents premature crepe hardening on the one hand while the incorporated methyl groups increase coating of the filler by the rubber polymer (so-called bound rubber) on the other. Secondary effects include a lowering of mixing viscosity, easier incorporation of the fillers, and efficient transmission of energy from polymer chains to the silica network, i.e. better tear-propagation strength.

$$\text{SiO}_2(\text{OH})_6 + (H_3C)_3-Si-NH-Si(CH_3)_3 \longrightarrow \text{SiO}_2(\text{OH})_5(\text{OSi}(CH_3)_3) + NH_3\uparrow$$

Silanes also augment the properties of organic elastomers. Particularly adept at this are mercaptosilanes for sulfur-curing and vinyl silanes for peroxide-curing.

Treatment of fillers

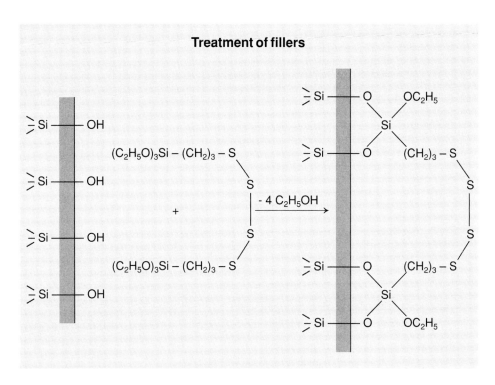

The effects achieved here vary greatly according to the types of filler and rubber. Particularly good results have been obtained in the following examples.

Rubber blend	Properties
EPDM with kaolin filler; additive: vinyl silane	Less susceptibility to moisture Faster extrusion Better mechanical and electrical poperties Increase in E modulus (around 60 %) Increase in tensile strength at maximum load (around 20 %)
SBR and natural rubber with quartz flour filler; additive: mercaptosilane	Less susceptibility to moisture Less heat build-up during vulcanization Less abrasion (tyres)

One minor disadvantage of silane additives is that they can affect the scorch time. The stability of elastomers towards reversion can be improved by a crosslinking combination of sulfur and silanes (e.g. tetrasulfane).

It is customary with organic elastomers to add the silanes in situ. Proprietary filler batches, e.g. quartz flour containing 50 % silane and 20 % plasticizer, are available for this purpose.

Silicone products additionally function as mould-release agents, e.g. in the manufacture of rubber gloves, and as lubricants, particularly for assembling EPDM window profiles.

The tyre industry imposes exacting demands on release agents.

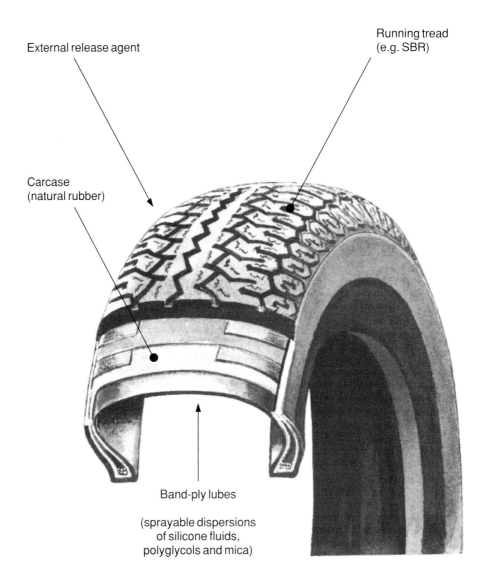

External release agent

Running tread (e.g. SBR)

Carcase (natural rubber)

Band-ply lubes

(sprayable dispersions of silicone fluids, polyglycols and mica)

The greatest problem here consists in preventing contact between the bladder (a rubber tube inflated with hot air) and the green tyre once curing is complete. The necessary separation is provided by special band-ply lubes, which are sprayable dispersions of silicone fluids, polyglycols and mica. They also prevent certain tyre defects, such as entrapped air and kinked beads, from occurring.

Silicone release agents in the rubber industry are complemented by silane adhesion promoters added to primers.

15. Textiles and leather

The main field of activity in the textiles industry is the production of fabric. The most important types of textiles are illustrated below.

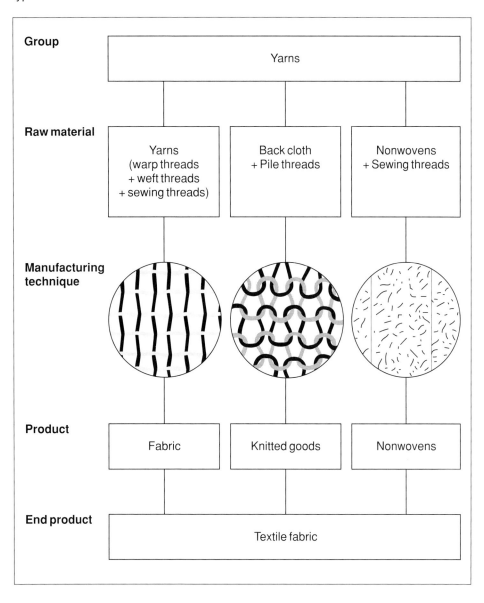

Yarns of the desired fineness (counts of 0.5 to $\geq$ 100 dtex) are produced by various techniques, such as worsted, carded-wool and open-end (OE) spinning. They are then usually knitted or woven to fabrics, which are differentiated according to the pattern by which the warp and weft threads are linked together, such as by linen, twill or atlas weaves.

Manufacturing, finishing and dyeing of textiles involves a great many different chemicals, the most important types of which are shown below. The predominant application area for silicones is that of high-grade finishing.

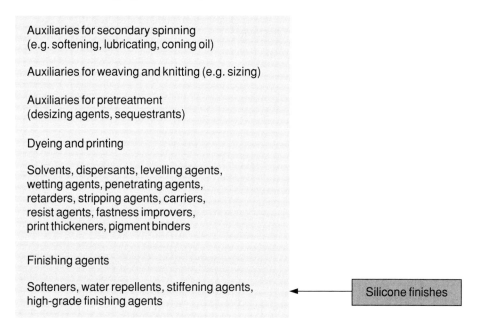

15.1. Silicone finishes

Silicone finishes fall basically into two categories, namely impregnations and coatings.

In the impregnation sector, silicone finishes function primarily as softeners. They can be rendered exceptionally hydrophobic for rainwear or hydrophilic for terry goods. In any event, silicones are valued particularly for the good handle which they impart. It is softer and fuller than that of organic softeners and, given appropriate recipes, can be made bulkier or silkier.

Silicone softeners can be used alone or in conjunction with a crosslinking agent. The type of crosslinking agent employed governs the degree of crosslinking (by condensation). Extensive crosslinking creates elastic finishes which primarily confer:

– soft handle and very good crease resistance on woven goods,
  and
– greater elasticity on knitted goods.

It is customary in the textiles industry to apply silicone finishes from aqueous phase, a fact which makes all low-add-on methods feasible, e.g. foam application.

The two most important methods are contrasted below.

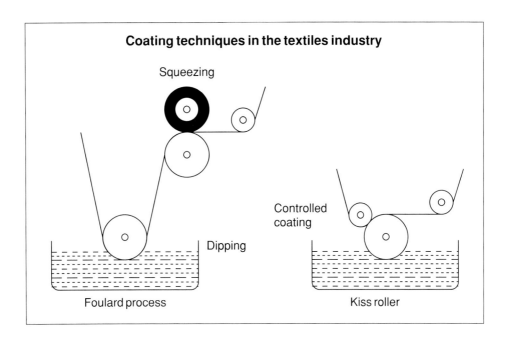

Foulard coating	Kiss-roller coating
Universal, widespread in textiles industry	Technique for minimum coating (liquor pick-up 25 – 35 %)
Liquor pick-up is about 60 %, excess is squeezed off	Advantages: much higher production rates; less energy required for drying

Softeners are not just restricted to high-grade finishing but can also be added to dye baths.

Unlike impregnated textile fabrics, which have an open structure, coated textiles have a sealed surface. The coatings are made from silicone elastomers, either alone or in combination with organic coating agents, such as acrylates, polyurethanes, and are characterized by interpenetrating networks.

The advantages of silicone coatings are that they are impervious to water but permeable to vapour.

Furthermore, they have a high thermal resistance (hence their use as substrates for ironing presses) and flame retardancy (use in vehicle seats), and are resistant to weathering and radiation (geotextiles for roofs of buildings). When used in conjunction with such organic coating agents as polyurethanes, silicones impart greater water imperviousness, increase tear strength and function as flow improvers and antiblocking agents.

The interest displayed by the textiles industry in silicone products has led to a number of special and in some cases very fashionable effects. Among these are stretch articles of clothing, which are usually denim, velvet and pile articles that have been impregnated with crosslinkable silicone finishes. The smooth or, alternatively, napped handle of extra-light-weight blousons is achieved with silicone impregnating agents, and so also is the chintz effect on velour. Silicone elastomer emulsions are used in the production of imitation suede leather with wrinkle recovery.

Similar applications are encountered in the leather industry, where silicones are also employed as water-repellent agents and softeners, e.g. silicone polyethylene dispersions. Silicone emulsions impart a distinctive sheen to furs.

Leather finishing is a very demanding area. A number of silicone additives are used to imbue leather-finishing materials (such as cellulose nitrate and polyurethane varnishes) with specific handle-modifying and flow properties. Silicones are also added to PUR coatings to yield specific handle properties.

To complete this section, let us turn our attention to some quite special finishing effects that can be obtained with silicone products.

Shrinking of wool in the wash can be successfully reduced through use of a silicone elastomer finish. In this application, the silicone coats the scaly woollen fibres, thus preventing them from matting together and shrinking.

The following silane has proved to be a highly effective antimicrobial finish for use on carpets, socks, etc.

$$(H_3CO)_3Si(CH_2)_3 \overset{CH_3}{\underset{CH_3}{-\overset{|}{\underset{|}{N}}^+}} - C_{18}H_{37} \, Cl^-$$

The biocidal action is not effected by the release of volatile agents.

Corsetry, garters etc. are coated with a silicone rubber antislip finish. The rubber used is of the RTV-1 type, which is readily applied by screen printing.

15.2. Dyeing and printing

The extreme temperatures and pressures involved in dyeing textiles create extensive turbulence and hence foaming. To be able to dye the textiles efficiently and uniformly, it is therefore necessary to add a highly effective antifoaming agent.

The following table shows the development of dyeing equipment down to the modern high-temperature, turbulent dyeing machines.

Development of dyeing machines

Dyeing conditions	Type of machine
Textile fabric at rest	HT beam dyeing
under mechanical motion	HT winch beck
Textile fabric in motion	Soft dyeing machine (overflow)
under hydraulic motion	HT jet dyer

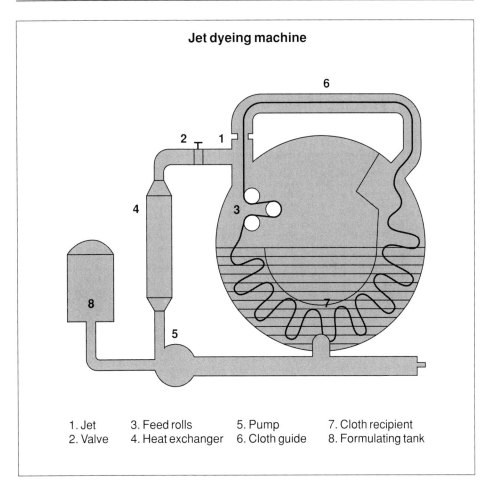

Jet dyeing machine

1. Jet
2. Valve
3. Feed rolls
4. Heat exchanger
5. Pump
6. Cloth guide
7. Cloth recipient
8. Formulating tank

Nowadays, the overwhelming majority of man-made fibres, especially polyesters, are dyed with disperse dyes. Dye liquors containing disperse dyes frequently contain large amounts of salts and carriers (adhesion promoters) that can adversely affect the stability of antifoam emulsions. In addition, the dyeing process involves severe heating and cooling phases during which the antifoams must continue to function. So-called self-emulsifying antifoams are preferred because they disperse so easily, a fact which guarantees uniform distribution throughout the dye liquor.

Apart from in dyeing, silicone antifoams find application in textile printing. Print pastes often contain added silicone fluids that impart a soft handle to the printed fabrics. Further bonuses are enhanced brilliance of the printing inks and better rub fastness.

15.3. Garments industry

The primary use of silicones in the garments industry is lubrication of sewing threads with silicone fluid finishes. Industrial sewing is a high-speed process, during which the thread heats up to about 300°C for a fraction of a second as it pierces the fabric. Conditions like these can cause the thread to break or melt, especially in the case of the predominant polyester yarns.

Optimum protection of polyester yarns against heat, and good lubricating action are afforded by soft finishes based on silicone fluids. These usually contain paraffin, which has good lubricating properties in the initial, cold phase. In the main, soft finishes are applied as the yarn is being made. Two methods are commonly used.

In the godet-roll technique, the silicone fluids are applied from small rolls.

In the exhaust process, the sewing thread absorbs the silicone fluid from a dilute liquor owing to special affinity. This step occurs in dyeing equipment after the dyeing stage. Under certain circumstances, it is possible to dye the sewing threads and to carry out finishing in one step; this is known as the all-in process.

Silicone release agents are required in many moulding processes for textile fabrication, especially of polyamide hosiery.

They are also used in the production of polyurethane shoe soles and have to fulfil requirements pertaining to overcoating and bondability of the demoulded soles.

Various RTV-2 silicone rubbers are employed in the manufacture of prototype shoes. Silicone gels have proved useful cushioning materials for sports shoes.

16. Paper industry

Silicones are used in the paper industry mainly for special (i.e. release) papers and in certain paper-printing processes. They are, however, also utilized to some extent in the manufacturing process.

16.1. Manufacture of paper

The manufacture of paper starts with the making of pulp and is illustrated schematically below.

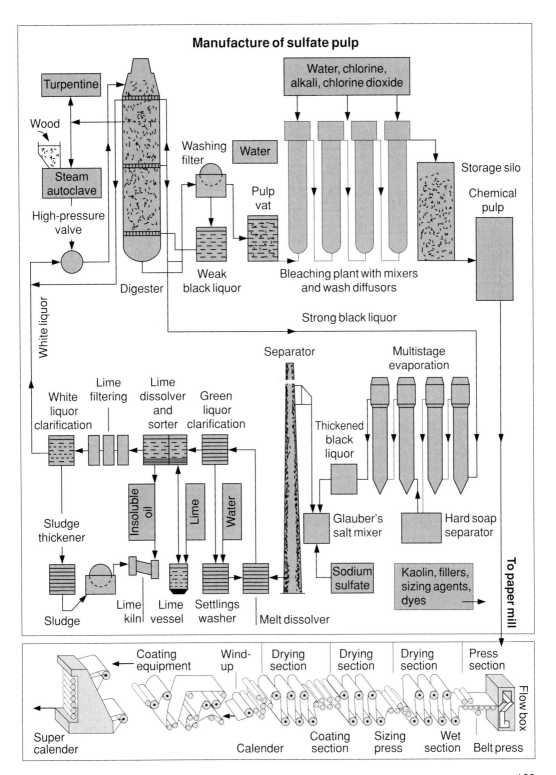

Two methods exist for making chemical wood pulp from ground wood pulp.

In the sulfite process, the lignin in the wood is converted to soluble ligninsulfonic acids by treatment with sodium hydrogen sulfite solution and sulfurous acid at 125 – 145°C. The resultant pulp is mainly used to manufacture fibres.

In the sulfate process, cooking is effected with a solution of sodium hydroxide and sodium sulfide. The reaction is carried out at 150 – 175°C and requires vast amounts of alkali (as much as 25 %), from which the products have to be recovered after the reaction is over. Pulp made by this process is primarily intended for the manufacture of paper. Antifoams are added during the cooking stage and as the black liquor is being evaporated. However, those usually employed contain only very minor amounts of silicones (0.2 %).

Pulp and added fillers are converted into paper in special calendering processes. Kaolin is a particularly important filler and is silanized to reduce the layer thickness of the paper and enhance printability.

A further opportunity to use antifoams presents itself at the paper-coating stage, which involves high processing speeds that generate a great deal of unwanted foam.

Special papers are made with the aid of silicone release emulsions. For instance, crepe paper is manufactured by passing the paper web over a rubber belt. A silicone emulsion sprayed onto the belt produces the micro-creping effect.

16.2. Silicone release papers

Coating paper with crosslinkable silicones affords silicone release papers that have a wide range of applications.

> Pressure-sensitive labels (ca. 70 %)
> for displaying prices, advertising, identifying products,
> labelling, electronic data processing etc.
>
> Industrial applications, e.g. transfer papers
> and backing papers (ca. 6 %)
>
> Polyethylene-coated laminated papers (ca. 5 %)
>
> Foodstuffs industry (ca. 7 %)
>
> Plastics industry (ca. 6 %),
> manufacture of laminates, films etc.
>
> Packaging (ca. 2 %)
> and adhesive tape manufacture (ca. 3 %)
>
> Miscellaneous (ca. 1 %),
> e.g. release papers for decorative films, magnetic tapes etc.

For reasons of price and quality, the silicone is applied as an ultra-thin, uniform coat. Consequently, particular attention is paid to the quality of the paper and the coating techniques employed.

Labels impose strict demands on the strength and dimensional stability of the paper and entail a special refining stage. Since the labelling process is critically dependent on the cutting process, it is best to use paper with a high content of short fibres. Preference is given to calendered paper, such as glassine and clay-coated paper, because they save on silicone. Paper that is too absorbent must be primed with poly(vinyl alcohol), methyl cellulose etc. in order to seal the surface and prevent the silicone from penetrating too far into the paper.

The amount of silicone applied can also be controlled via the coating system. In the most common of all gravure coating techniques, i.e. reverse coating, this is effected by the use of half-tone rollers. However, other coating systems, consisting of several rubber and steel rollers, have become popular for solvent-free silicone products.

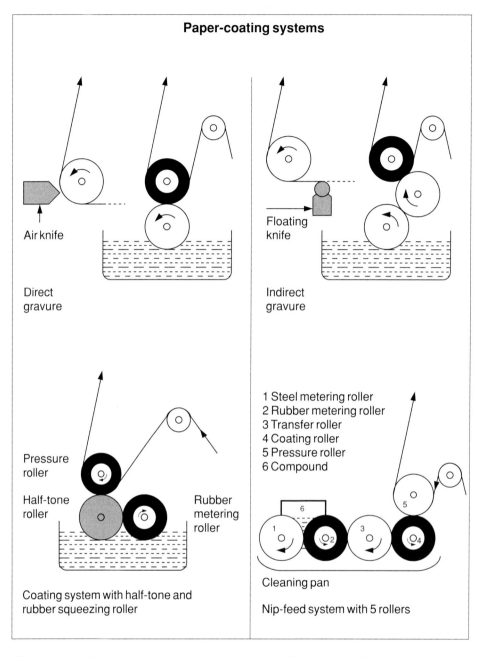

Paper-coating systems

Direct gravure — Air knife

Indirect gravure — Floating knife

Coating system with half-tone and rubber squeezing roller
- Pressure roller
- Half-tone roller
- Rubber metering roller

Nip-feed system with 5 rollers
1 Steel metering roller
2 Rubber metering roller
3 Transfer roller
4 Coating roller
5 Pressure roller
6 Compound

Cleaning pan

Silicones and adhesives can be applied in two steps, with time being allowed for the silicone to fully cure. This is known as the off-line technique. Alternatively, the adhesives can be applied immediately after the paper web has been coated with the silicones. This is the in-line process and imposes very high demands on the curing rate of the silicones.

The very sparse silicone coats (0.5 g/m^2) cure very rapidly by either addition or condensation at temperatures of 100–200°C.

Since the coatings are so thin, the paper can also be siliconized photochemically and by radiation. Systems have already been developed for the following crosslinking mechanisms.

– UV-IR crosslinking (hot curing)
  Crosslinking follows a hydrosilylation reaction and involves a special noble-metal catalyst, which functions both as a UV sensitizer and a temperature-sensitive IR catalyst.

– UV crosslinking (cold curing)
  Photosensitizers are used to promote free-radical addition.

– Electron-beam crosslinking (EB curing)
  The high energy of electron-beam curing renders the use of sensitizers superfluous. However, a nitrogen atmosphere is required because otherwise oxygen radicals would form and interfere with polymerization.

Radiation curing is becoming increasingly popular for many silicone applications. The diagram below illustrates the most important applications after manufacture of release papers.

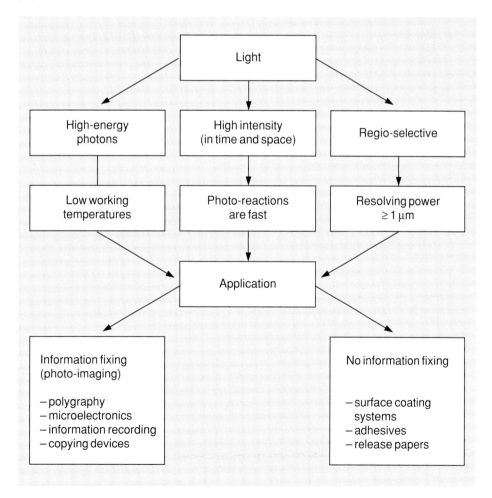

The main criteria to be adopted in assessing release papers are the values and constancy of the release force required, the resistance of the silicone film to rub-off and the degree of curing, expressed in terms of subsequent adhesion.

Values for release force depend firstly on the rate at which the release paper is pulled from the adhesive layer. Differences can arise according to whether the release paper is removed slowly or quickly. Speeds of 100 – 250 m/min are often employed in mechanical labelling.

Of greater importance is the type of adhesive in contact with the release paper. Conventional adhesives have an elastomeric or resinous base and contain plasticizers, fillers and stabilizers. In the case of silicone release papers, the tendency is away from resinous adhesives, e.g. acrylates, and towards their elastomeric counterparts, e.g. SBR, polybutadiene. The former contain numerous double bonds that can react with the H-siloxane crosslinking agent and thus impair the value and constancy of the release forces. Added to which, uncrosslinked silicone polymers could migrate into the adhesive, thereby reducing its adhesiveness.

Good rub-off resistance of silicone films depends on good anchorage to the paper and on thorough curing. Here, the mechanical properties of the silicone rubber film also play a role.

Bilaterally siliconized release papers permit the adhesive to be transferred from the paper direct to a substrate, provided that the two silicone films differ in their release force. This is known as the controlled release effect.

The mono-web process dispenses with the need for silicone release papers because the labels are printed first, siliconized on the printed sheet and then coated with adhesive on the reverse side.

Release papers for gelling processes (cf. Plastics industry) must be resistant to high temperatures in addition to possessing release properties. These papers are pre-coated several times with thermally resistant coatings and a water barrier in order to obtain blister-free removal of the cast films (e.g. Warren papers).

16.3. Printing techniques

Various printing techniques utilize silicones because, as a rule, the binders in the inks tend to foam. Consequently, antifoams have to be added. In the cases of relief and intaglio printing, modified acrylate binders are used. The flow properties of the ink are particularly crucial in screen printing. They are regulated by the addition of fumed silica, which also prevents the pigments from settling.

The greatest use of silicone occurs in the very important and widespread offset printing technique. Offset printing is a planographic technique in which the printing, oleophilic areas and the non-printing hydrophilic areas (films of water) must be sharply delineated. Addition of hydrophobic fumed silica to the printing ink prevents water absorption at the interfaces and the formation of an emulsion, which would lead to blurred printing.

Printing at high speeds can also cause blurring. If the ink has not fully dried, it smudges over the rollers. In modern offset printing machines, it is usual to add a small quantity of silicone fluid emulsion to the spray used to moisten the paper after the drying stage. The fluid acts as a release agent and lubricant on the rollers, preventing the moist, plasticized printing paste from adhering to the rollers. It inhibits the formation of nibs as the paper webs are pulled over the rollers and it enhances the gloss and smoothness of the print.

Paper is not the sole substrate to be printed with the aid of silicones. Uneven surfaces, such as glass bottles, ceramics and plastics articles are printed by various, specially developed, flexographic techniques. In the tampo printing technique, rubber-elastic, soft tampons transfer the ink from the printer's copy to the article to be printed. The tampon absorbs the ink from the block or template and faithfully reproduces it on the various surfaces of the article.

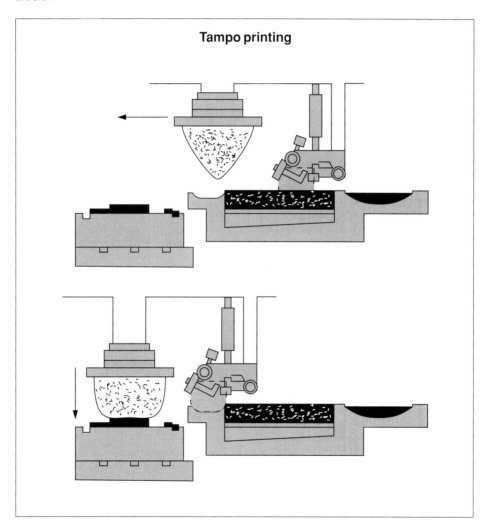

Tampo printing

A variant of the tampo printing technique is the hot-stamp process in which a printing film is transferred from a stamp that is coated with silicone rubber.

Silicone rubber has proved to be a highly useful material for making printing tampons. Unlike other products, such as gelatin, it possesses exactly the right release properties and its great resilience allows rapid print runs.

17. Construction industry

The construction industry is the greatest user of silicone products. This is mainly due to the use of silicone jointing materials, which have gained an undisputed leading position in construction technology.

17.1. Joint sealants

Masonry is constantly subjected to movement and stresses, the causes of which may be temperature changes, moisture, shrinkage of construction materials, subsidence of the substrate etc. The gaps produced by such movements can only be filled by an elastic jointing compound.

In choosing a suitable jointing material, it is therefore best to evaluate its deformation and mechanical properties, and to determine the movement to be accommodated by the joint. Values for these properties can differ enormously according to whether the material is elastic or plastic.

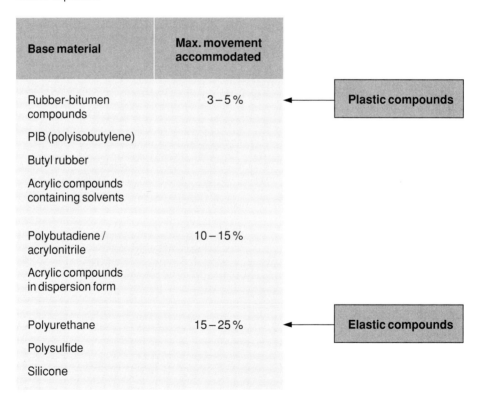

Base material	Max. movement accommodated	
Rubber-bitumen compounds	3–5 %	← Plastic compounds
PIB (polyisobutylene)		
Butyl rubber		
Acrylic compounds containing solvents		
Polybutadiene / acrylonitrile	10–15 %	
Acrylic compounds in dispersion form		
Polyurethane	15–25 %	← Elastic compounds
Polysulfide		
Silicone		

Differences in deformation behaviour are most obvious from stress-strain diagrams. Unlike purely plastic behaviour, rubber-elastic behaviour results in an **increase in stress** with progressive strain.

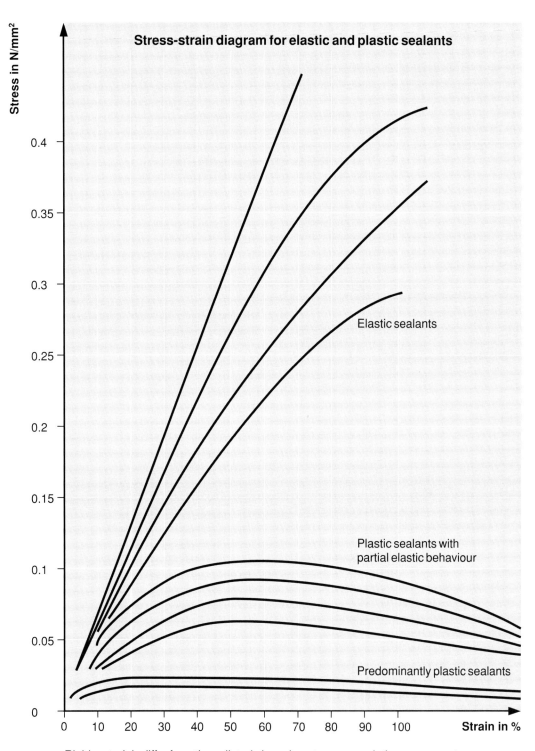
Rigid materials differ from those listed above in not accommodating movement.

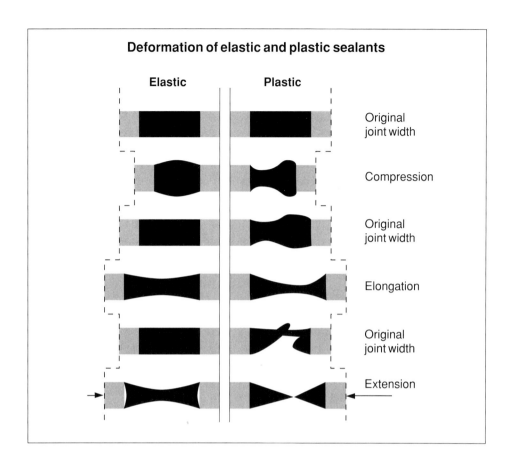

The degree of plasticity or elasticity is the prime factor governing the use profile of a jointing material.

Highly plastic compounds that are subject to repeated elongation and compression (so-called chewing gum effect) eventually suffer cohesion failure. Very elastic compounds in the extended state exert a strong tensile force on the edges of the joint owing to conserved strain. This can result in adhesion failure if, for example, the joint is too thin or the primer is not effective enough.

For movement joints, behaviour between elastic and plastic is desirable. In the extended state, materials of this kind (plasto-elastic or elasto-plastic types) undergo a gradual decrease in stress called relaxation, as a result of which the tensile force exerted on the joint edges diminishes.

Some idea of the extent of relaxation is given by the following comparative values for the recovery of jointing materials.

Type of rubber	Recovery after 24 h	
Elastic	70–100 %	(for normal types of
Elasto-plastic	40– 78 %	silicone rubber 80 – 100 %)
Plasto-elastic	5– 40 %	
Plastic	0– 5 %	

On account of the demands listed above, materials used for movement joints are of the elasto-plastic type with a degree of recovery of 60 – 70 %.

A comparison with organic jointing materials reveals that silicones are better at accommodating movement.

Sealant	Curing mechanism	Movement accommodated
Modified materials (natural oils, synthetic resins)	Oxidative surface drying	2 – 5 %
Solvent-borne acrylates Polycarboxylates	Evaporation of solvent	5 – 8 %
Acrylic dispersion	Evaporation of water	10 – 20 %
– Polysiloxanes – Polysulfides – Polyurethanes	Atmospheric hydrolysis of crosslinking agent (one-component systems)	20 – 25 % 15 – 25 % 15 – 25 %
– Polysiloxanes – Polysulfides – Polyurethanes	Addition of crosslinking agent (two-component systems)	10 – 25 % 10 – 25 %

The greatest advantage of silicone jointing materials is their extreme durability, which stems from their stability towards weathering, radiation and chemicals. Moreover, silicone jointing materials are safe and easy to apply and can be readily formulated.

The RTV-1 silicone rubbers used as jointing materials are classified according to the curing mechanisms involved. The most important of these are shown below.

Crosslinking of RTV-1 silicone rubber

$$HO-\underset{\underset{R}{|}}{\overset{\overset{R}{|}}{Si}}\sim O-\underset{\underset{R}{|}}{\overset{\overset{R}{|}}{Si}}-OH + 2\,R-SiX_3 \longrightarrow$$

$$R-\underset{\underset{X}{|}}{\overset{\overset{X}{|}}{Si}}-O-\underset{\underset{R}{|}}{\overset{\overset{R}{|}}{Si}}\sim O-\underset{\underset{R}{|}}{\overset{\overset{R}{|}}{Si}}-O-\underset{\underset{X}{|}}{\overset{\overset{X}{|}}{Si}}-R + 2\,HX$$

(further hydrolysis/condensation steps with H$_2$O yielding crosslinked network + 4 HX)

Classification of crosslinking systems employed in one-component silicone sealants

Type	System	Characteristic group (X)
Acidic	Acetoxy	$CH_3-CO-O-$
	Octoate	$R-CO-O-$
Neutral	Amide	$R-CO-NR'-$
	Oxime	$R_2C=NO-$
	Alkoxy	$RO-$
Alkaline	Amine	$R-NH-$

These groups of products will be soon complemented by pressure-sensitive two-component systems which are still being developed for use in the construction industry.

Each curing mechanism confers its own distinctive properties on the various systems. Consequently each system has its own specific field of application.

Alkaline systems	Good adhesion to construction materials, glass, ceramics
	Ready compounding (amine system)
	Good mechanical properties
	Environmental stress cracking of polycarbonates
Neutral systems	No odour evolved during crosslinking
	No corrosion of metals or plastics
Acidic systems	Good adhesion to glass, ceramics, metals and plastics
	Excellent transparency (acetoxy system)
	Good mechanical properties
	Environmental stress cracking of polyacrylates

The properties of silicone jointing compounds are greatly affected by their constituents, the most important of which are shown below.

Formulation of sealants

Constituent	Function
Polymer	Binders for forming elastic networks
Plasticizers	Adjustment of mechanical properties, such as hardness, elasticity etc.
Reinforcing fillers	Thixotropy (non-sag), cohesion (adjustment of mechanical properties)
Non-reinforcing fillers	Reduction in costs, adjusting of flow properties, hardness, flame retardancy
Additives	Adhesion promotion, ageing resistance, altering of curing rate
Crosslinking agents	Crosslinking of polymer components

The consistency can be varied extensively by altering the proportions of fillers, polymers and plasticizers. Of the numerous proprietary jointing materials on the market, those possessing the following properties stand out the most.

– Extremely soft, elastic materials for use in expansion joints
– Materials with very rapid or slow (delayed) skinning time
– Materials with high mechanical strength and resistance to chemicals
– Resilient and non-sag materials

A critical factor in many jointing applications is the "non-sag" property of the material, i.e. the resistance of the material to flow after application to a vertical surface. It is achieved by adding fumed silica, the most important filler, to the jointing compounds. The effect is thixotropic and due to the special agglomeration behaviour of the silica particles. When they are at rest, they build up a silica network via the formation of hydrogen bonds. Under the action of shear forces, the network breaks down readily, e.g. as when pressure is exerted on a tube, but rapidly reforms when the pressure is removed. Thus, the material does not flow.

The range of available silicone joint sealants is so comprehensive that they are able to perform a multitude of tasks in the construction industry.

Expansion joints on building facades are usually sealed with materials of low elastic modulus. Such materials can accommodate up to 50 % movement in both directions, to satisfy the various building standards. In this sector, the jointing materials must adhere well to alkaline building substrates, such as concrete, without the need for a primer. Modified amine systems satisfy this need.

Materials employed for sanitary joints, such as those between bathroom tiles, must have good abrasion resistance . Both amine and acetoxy types can be used here.

Connecting joints for windows with frames of aluminium or plastic (PVC) are generally made from silicone materials. The advantages are excellent adhesion, great resistance to weathering and good transparency, especially in the case of acetoxy types. For the same reasons, silicone rubbers are used for aquaria.

Silicones have also found their way into the manufacture of insulating glass. Double-glazed windows are usually constructed in the following way. Once the aluminium spacer has been applied, a seal of butyl rubber is inserted to act as an effective barrier against atmospheric moisture and water vapour. This seal is itself sealed with a curable liquid, for which application silicones have the advantages of easy handling and brief skinning times.

In road construction and in airports, silicone jointing materials known as highway sealants serve to create elastic joints between the concrete slabs.

Prefabricated profiles made of HTV rubber are another type of seal found in the construction industry and are used to insulate windows and doors. Compared with their organic counterparts, they are more flame retardant and highly functional. The latter property manifests itself in the ease with which such windows and doors can be closed.

Dispersions of one-component rubbers make useful horizontal seals for the building industry. Main applications here are the coating of wells, channels and potable-water tanks.

Modified silicone rubbers, such as combinations with bituminous materials, are now finding use in roofing systems.

The use of silicone rubber as a mould-making material has made it possible to design structured facades of the most diverse shape and material, e.g. from concrete. This is accomplished by the shell technique, which permits manufacture of very large sheets with textured surfaces. Methods employing silicone rubber dies have made a decisive contribution towards the spread of structured facades and injected life into modern building design.

In structural glazing, the silicone compounds constitute the only supports between the glass sections. This technique allows large facades to be built purely of glass sections without the need for metal retainers and it has enriched architecture with new and impressive designs.

Geotextiles provide new opportunity for designing fabric roofs. Silicone rubber is ideal for this application on account of its excellent weathering resistance. It does not yellow or become embrittled, nor is it dirtied by dust deposits, provided that the surface is additionally impregnated with silicone resin.

17.2. Masonry protection

Silicones are not just important as rubber jointing compounds. They also protect and preserve virtually all major building materials.

Building materials can be attacked in various ways but absorption of water is always involved.

They have a more or less porous structure, which manifests itself as channels, blisters and cavities. Clogging of such voids with water diminishes the thermal insulating properties and all sorts of damage can occur.

Water can be absorbed by any of a number of mechanisms.

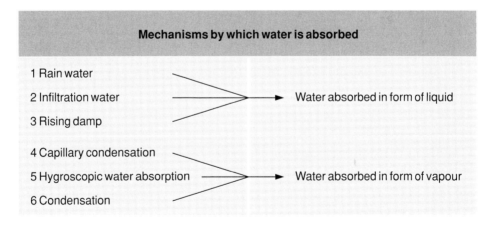

Irrespective of the absorption mechanism, the masonry becomes thoroughly damp, allowing weathering to take effect.

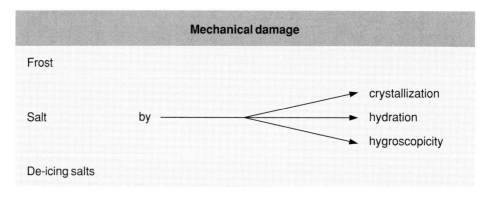

Mechanical damage		
Frost		
Salt	by	crystallization / hydration / hygroscopicity
De-icing salts		

Chemical corrosion				
Loss of binder through chemical conversion (formation of salts)				
$CaCO_3$	$+ H_2SO_4$	→	$CaSO_4$	$+ H_2O + CO_2$
Lime	(Acid rain)		Gypsum	Carbon dioxide
Insoluble binder			Slightly soluble salt	

Biological corrosion	
Bacterial and fungal attack	Salt formation through metabolites

In order to prevent the materials from being damaged, it is imperative that the water does not succeed in penetrating the masonry. There are two fundamentally different ways of accomplishing this, namely, by applying film-forming coatings or hydrophobic impregnating agents that leave the pores open.

Film-forming surface coatings include the following types.

– Colourless systems of highly concentrated solutions of resins. These are extremely effective sealants and can withstand water under pressure.

– Coloured (pigmented) systems based on film-forming paints. This group includes emulsions and synthetic resin paints as well as synthetic resin plasters that contain pigments and fillers. The degree of impermeability to water conferred depends on the type and proportion of the synthetic resin.

Hydrophobic systems form hydrophobic, open-pore, molecular films that prevent the water from penetrating. The surface of the building material becomes water repellent without losing its permeability to water vapour. Products in this group include silicones and stearates.

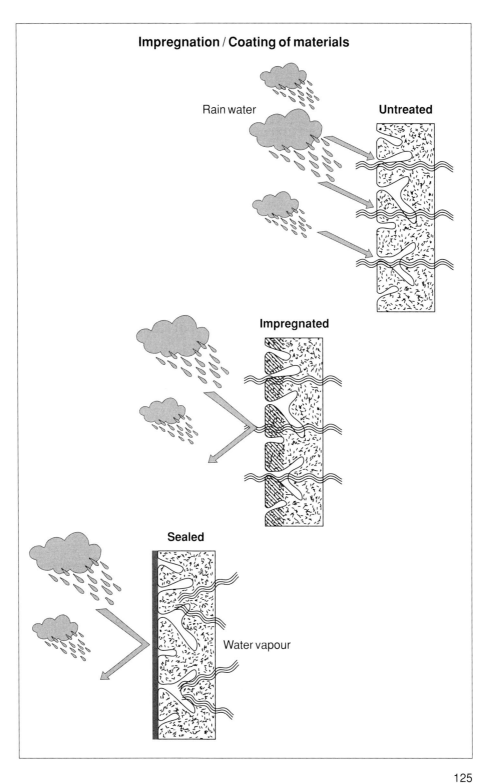

The most important products are shown below, along with the form in which they are used and a description of their properties.

Limewashes are not used much nowadays because of their very poor weathering resistance.

Pigmented coating systems

Coating	Binder	Solvent and diluent	Properties
Pure silicate paint	Potassium silicate (potassium waterglass)	Water	Two-pack systems with excellent vapour permeability. Protection against moisture is relatively low. One disadvantage is salt formation. $K_2SiO_3 + CO_2 + H_2O \longrightarrow SiO_2 + K_2CO_3$
Silicate emulsion paints	Potassium silicate Synthetic resins	Water	One-pack system, similar to pure silicate paint
Emulsion paints	Synthetic resins, e.g. poly(vinyl-acetate), styrene, acrylate	Water	Most common coating system with good protection against moisture; disadvantage is low water vapour permeability
Polymer paint	E.g. acrylic resin	Organic solvent	Impermeable to running water and water vapour
Silicone resin paint	Silicone resin emulsion	Water	Pigmented coating system with good water vapour permeability and weathering resistance
Siloxane paint	Silicone resin / acrylic resin	Organic solvent	Similar to polymer paint

Colourless coating systems (impregnating agents)

Impregnating agent	Active ingredient	Properties (reduction in water vapour permeability in %)
Siloxane resin	Methyl polysiloxane	5 – 8
Siloxanes	Low-molecular, highly alkylated polysiloxanes	5 – 8
Silanes	Higher alkylated silanes e.g. isooctyl triethoxy silane	5 – 8
Silicates with hydrophobic additives	Alkali silicates (water-glass), alkali siliconates	5 – 10
Silicic acid esters with hydrophobic additives	Tetraethyl silicate/silane	15 – 30
Acrylic resins	Polymethacrylates	15 – 25
Metallic soaps	Stearates of aluminium and titanium	5 – 15

The following diagram shows that reducing absorbency is only one aspect of masonry protection. It is equally important for the treated building materials to dry out thoroughly. This means that they must be very permeable to water vapour.

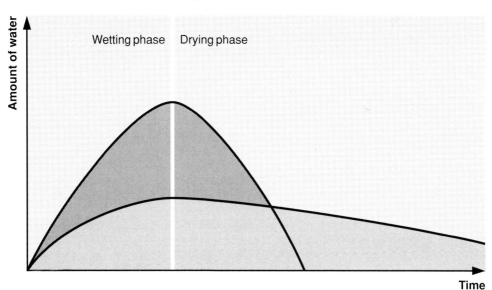

Silicone masonry water repellents offer the following advantages over similar agents.

- Much lower water absorption
- Retention of permeability to water vapour
- Deep penetration
- Durability
- No gloss
- No yellowing on exposure to weathering
- No dirt pick-up, i.e. tack-free drying
- Resistance to UV radiation

All of the different classes of products available for treating building materials have one feature in common, namely they deposit a very thin film of silicone resin.

A few considerations have to be borne in mind if optimum effects are to be achieved. First, the silicone impregnating agent must penetrate sufficiently far into the material. But, as the following table shows, building materials are not alike in their absorbency.

Minimum depth of penetration and consumption of impregnating agents for different building materials

Material	Minimum depth of penetration (mm)	Mean consumption (l/m^2)
Asbestos cement	1	0.1
Fine-faced concrete	1–2	0.2–0.4
Aerated concrete	6–8	1.0
Lime sandstone	3	0.4
Mineral plaster	3–4	0.5
Sandstone, dense	2–3	0.3
Sandstone, absorbent	5–6	0.7–1.0
Brick, dense	2–3	0.4–0.6
Brick, absorbent	6–8	1.0–1.5

The second factor governing the durability of a silicone impregnation is the alkalinity of the building material.

Alkalinity of major building materials

Building material	Age	pH
Asbestos cement	Fresh 6 – 12 months old	12.0 – 13.0 11.0
Concrete	Fresh 6 – 12 months old	11.0 – 12.0 9.5 – 10.5
Jointing mortar (based on cement)	Fresh 6 – 12 months old	10.5 – 11.5 9.0 – 10.0
Aerated concrete	Fresh 6 – 12 months old	8.5 – 9.5 8.0 – 8.5
Lime sandstone	Fresh 6 – 12 months old	9.0 8.5
Lime-cement plaster	Fresh 6 – 12 months old	10.0 – 11.0 9.0 – 10.0
Sandstone	–	7.0 – 8.0
Facing brick	–	7.0 – 8.0

Alkali-resistant silicone water repellents have been developed for use with highly alkaline building materials, such as concrete.

A further advantage of silicone water repellents is their sheer versatility. They can be used for impregnating a wide variety of building materials, such as natural stone (with silicates or carbonates as binders), ceramics, roof tiles, cement-bound materials (cement, asbestos cement, cement-bound plasters), lime-bound materials (aerated concrete, lime sandstone, lime plaster) and gypsum.

Other applications within the construction industry include:

– Facade impregnation and restoration (by flooding, brushing and spraying)
– Damp-proofing of walls (by injection)
– Preservation and restoration of natural stone
– In-plant impregnation of building materials (by dipping, admixing)
– Additives for mineral plasters and gypsum
– Primers and pigment additives

These include primers for silicone-resin and emulsion paints. They also embrace silicone-resin dispersions, which are binders for silicone-resin paints but serve additionally as water repellent additives for other paints (e.g. limewashes and one-pack silicate paints).

Possible applications as primers or impregnating agents for coating systems are summarized in the diagram below.

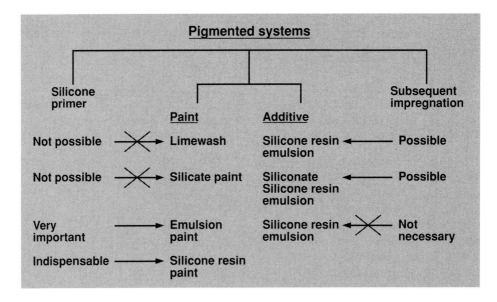

Note that the use of a silicone primer for emulsion paints is highly recommended. However, emulsion paints cannot be protected subsequently by a silicone impregnation.

The following examples are intended to demonstrate that silicones are just as useful for preserving old buildings as they are for protecting new ones.

Chemical corrosion of old buildings and monuments is making it increasingly more imperative to adopt measures to preserve the stone. Excellent results in this respect can be obtained with tetraethyl silicates, which, on hydrolysis, generate silica that fills up and strengthens the pores of decomposed materials.

$$Si(OC_2H_5)_4 \xrightarrow{H_2O} SiO_2 + C_2H_5OH\uparrow$$

The salient point here is that the stone is strengthened by a substance of similar composition, namely silica, and not by a different material.

Since concrete is the predominant construction material nowadays, measures also have to be adopted to protect it.

Concrete located in the vicinity of traffic, e.g. bridges, multistorey car parks etc., comes under severe attack from de-icing salt. The results are frost damage and corrosion of the embedded steel reinforcements. Normally, access of corrosive substances (H_2CO_3) and rusting of the reinforcements are prevented by the large reserves of $Ca(OH)_2$, which forms as the concrete hardens.

Cement

$$\begin{pmatrix} CaSiO_4 \\ Ca_3Al_2O_6 \end{pmatrix} + H_2O \longrightarrow \begin{matrix} Ca(OH)_2 \\ Al(OH)_3 \end{matrix} \longrightarrow \text{Intermatting to closely packed crystals in the hardened concrete.}$$

The bursting forces generated by salts and rusting reinforcements can destroy the tightly packed, matted structure of the cured concrete. Therefore, in order to prevent penetration by the dissolved salts, it must be impregnated with silicones. Silanes and siloxane mixtures are particularly effective for this purpose and have the advantage of being able to penetrate far into dense materials. Products containing them are very resistant to alkalis and play a decisive role in enhancing the resistance of concrete to freezing and to de-icing salt.

Lightweight building materials, which are very prevalent in modern construction techniques, differ from highly compact concrete in being extremely water absorbent. Hydrothermally hardened materials, such as aerated concrete and lime sandstone, are surface-impregnated or impregnated with silicones in the manufacturing plant immediately after production. As time passes, the protection afforded by the impregnation allows the material to dry out and lose its water content of around 30 %.

17.3. Insulating materials

It is in the impregnation of lightweight materials that the thermal insulation properties of silicones really come to the fore.

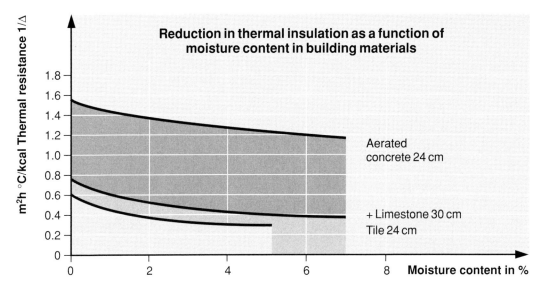

The diagram above clearly demonstrates the extent to which the thermal resistance of porous materials depends on the moisture content.

As the diagram below shows, cellular materials are particularly good thermal insulators.

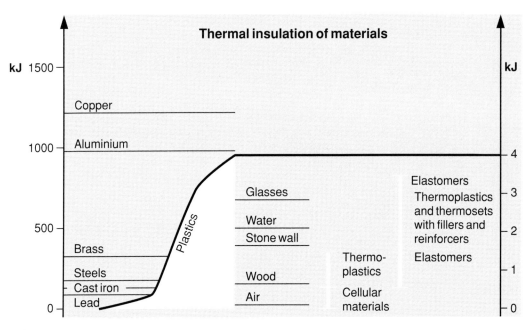

The building industry utilizes a number of cellular materials for insulation, e.g. cellular foams, mineral-fibre insulators containing synthetic resins as binders, and loose fillers.

Mineral-fibre insulators made of glass wool or rock wool, mainly containing phenolic resins as binders, are often treated with silicone fluids or resins in order to reduce their water absorbency.

Loose fillers are expanded powders such as perlite, vermiculite and aerated-concrete granules that are kept dry by means of silicone impregnating agents. The impregnating agent is sprayed on during production, e.g. in the case of perlite, at a temperature in the range 400 – 450°C.

Insulating materials based on resin foams, such as expanded PUR, can be coated with silicone to prevent their absorbing water as in PUR roof insulation. Insulating coatings can also be made by spraying on blown glass wool coated with silicone impregnating agents. District heating pipes benefit from insulation based on silicone-coated aluminium foil. Foamed silicone rubbers are worth mentioning because of their flame retardancy. They are used for applications in which priority is attached to safety.

18. Surface coating compounds

Coatings are film-forming agents that are characterized by hardness, brilliance and weathering resistance. Their function is to protect surfaces against damage, thermal stress, weathering and corrosion and to impart a decorative appearance. The most important constituents of surface coating compounds are as follows.

– Binders (resins, plasticizers and film-formers)
– Pigments, fillers and special additives for thixotropic effects, antisettling agents etc. Unpigmented coatings are known as varnishes
– Solvents (organic or aqueous), extenders to improve brushability

Coatings can also be used in the form of powders, in which case they are entirely free of solvents. Coatings with a low proportion of solvent are known as high-solids (solids content lies between 70 and 90 %). Water-borne coatings constitute a special group and are increasingly being used for industrial mass coating by the ELT technique.

The properties of coatings depend critically on both film-forming power and curing. During the latter stage, the binders must not separate from the pigments. Coatings are classified according to their curing mechanism.

– Air-drying coatings (oleous types such as alkyd finishes and oil-free types such as cellulose nitrate)
– Stoving enamels (such as silicone paints, alkyd-melamine resin coatings)
– Reactive coatings (one-pack and two-pack types, such as ethyl silicates, epoxy and polyurethane paints)
– Radiation-curing coatings (e.g. acrylic resin paints)

Silicone paints are primarily stoving finishes that require temperatures of around 200°C to effect curing because room temperature will only cause physical drying. Drying can be accelerated by catalysts, e.g. metallic salts, or blends with butyl titanate.

$$-\underset{|}{\overset{|}{Si}} - OH + HO - \underset{|}{\overset{|}{Ti}} - \longrightarrow -\underset{|}{\overset{|}{Si}} - O - \underset{|}{\overset{|}{Ti}} -$$

However, it harbours the risk of rapid embrittlement at elevated temperatures.

Coatings can be applied by a number of techniques, including spraying, pouring, dipping, rolling, electrophoresis, and electrostatics. One special technique is that of coil coating, which is most often used for silicone resin blends, particularly silicone polyester paints. Surface coatings are usually applied in several layers:

– Primer, e.g. shop primer, wash primer

– Filler

– Hard, brilliant top coat

The application areas of surface coatings and the corresponding uses of silicones are shown below.

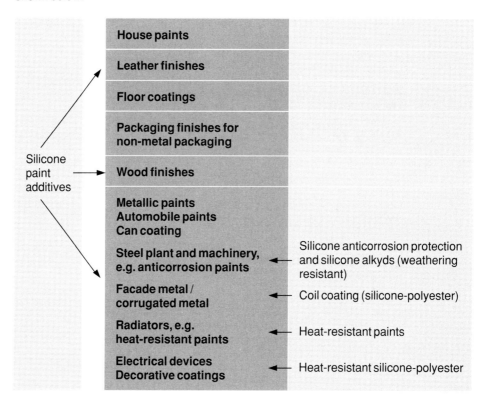

The greatest demand is for metallic paints, which also constitute the most important usage of silicones.

The main uses of polysiloxanes are as follows.

– Methyl and methylphenyl resins: for heat-resistant paints

– Phenylmethyl alkoxysiloxanes (silicone-intermediates): made to react with organic resins

– Silicone resin blends: for coil coating and as decorative finishes

– Ethyl silicates: for anticorrosion paints

Silicone intermediates provide great scope for performing a large variety of reactions. Since they are methoxy and hydroxy functional oligomers, they are capable of reacting with organic resins.

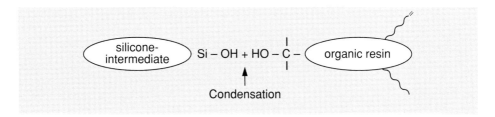

The best co-reagents are polyester resins, which are made from polyhydric alcohols and polybasic acids, and alkyd resins, which contain unsaturated acids known as fatty oils. These reagents can be simply mixed in the cold to yield resin blends.

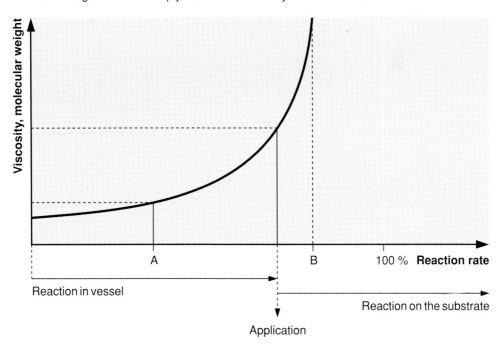

Optimum properties, such as weathering resistance, are only obtained by making the reagents condense with each other. To this end, they are boiled together for several hours. The extent of the reaction is followed by measuring the increase in viscosity. The reaction must be interrupted before gelling occurs. Since the components differ in reactivity, it is important to choose the correct conditions for the reaction.

Combining silicones with organic resins has several advantages that are exploited in the following applications.

- The silicone component enhances the weathering resistance, gloss retention, dirt repellency and anti-chalking properties of the organic resins.

- The organic component confers readier curing and greater hardness on the coatings. In addition, the coatings are less susceptible to attack by organic solvents.

18.1. Heat-resistant paints

The diagram below indicates the suitability of methyl and phenyl resins for heat-resistant pigments for use at different temperatures.

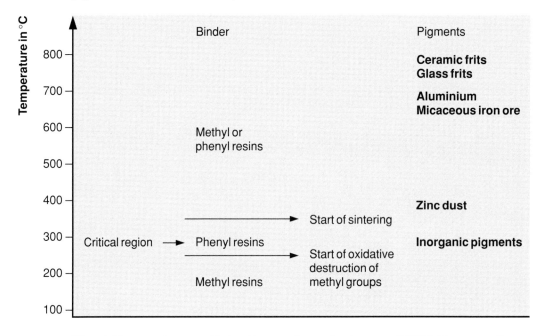

Phenylsilicone resins can withstand temperatures of up to around 300°C. Consequently, low-pigmented, high-gloss paints can be used up to these temperatures. The temperature range of 250 – 350°C is critical in as far as oxidative degradation of methyl groups occurs from 250°C on and sinter processes first occur from about 350°C on. Above 350°C, the choice of silicone resin for use as binder is irrelevant. Sintering of aluminium pigments produces highly adhesive coatings that consist of aluminium silicates. Paints pigmented with aluminium afford excellent anticorrosion protection at elevated temperatures that is due to the leafing effect of the aluminium.

So-called decorative paints for coating electrical devices, cooking utensils (enamel substitute) are often based on silicone polyester resins, with the silicone content being in the range 50 – 80 %.

Application areas for heat-resistant silicone paints include ovens and cooker hobs (micaceous iron paints), exhaust-manifold paints for automobiles, aluminium paint for protecting equipment, and coating of incandescent light bulbs.

Epoxy resins containing 30 – 50 % silicone are especially suitable for the electrical sector.

18.2. Coatings with greater weathering resistance and flexibility

One application of highly flexible, rapid-curing coatings is that of coil coating, in which metal is coated before it is shaped.

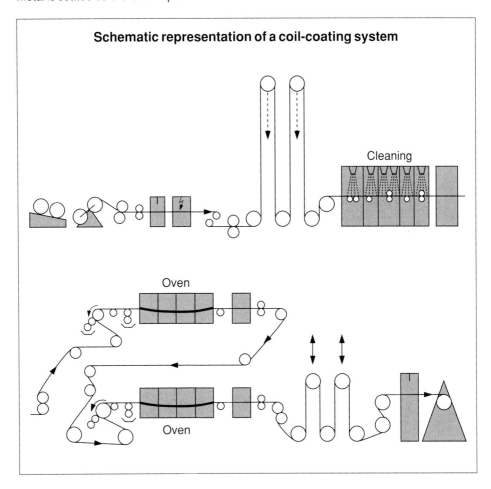

Schematic representation of a coil-coating system

Predominant in coil coating are silicone polyester coatings with a silicone content of 15 – 50% and silicone acrylic coatings, with a silicone content of 10 – 30%. These systems are characterized by high reactivity (curing takes place within 60 s at 250°C), good mechanical properties (flexibility, impact resistance) and good weathering resistance (colour retention, gloss). Competing systems for coil coating are acrylic paints, pure polyester resins and, for very exacting demands, fluoropolymers.

Outstanding weathering resistance is expected of coatings for marine objects and manufacturing plant. Temperatures in these applications are much lower, lying in the range of 100 – 150°C. The requisite properties are provided by silicone-modified alkyd systems that contain about 30 % silicone.

18.3. Anticorrosion coatings

Excellent anticorrosion coatings can be made from ethyl silicates and zinc dust. These types are used as

– shop primers (one-pack system),
  wash primers (two-pack systems), and
– top coats.

It is also possible to combine them with organic components, such as poly(vinyl butyral) and acrylates.

Ethyl silicates offer a number of advantages over anticorrosion coatings based on epoxy resins or chlorinated rubber, namely

– high resistance to chemicals, oils and solvents;
– thin, elastic films that are easy to paint over and are ideal for shop and welding primers;
– retention of anticorrosion protection even when the binder has been damaged mechanically (e.g. through formation of local cells).

These advantages, however, only come to bear when the substrate is clean and has been well pre-treated, i.e. sandblasted. Rusty substrates should be coated with such products as red lead.

Anticorrosion coatings based on ethyl silicate are mostly applied to steel constructions, underwater equipment, ships, storage tanks for fuel oil, pipelines, water-treatment and purification plants.

18.4. Coatings additives

Optimum coating properties depend not only on the binder but also to a great extent on the pigments, fillers and auxiliaries. Pigments, particularly TiO_2, can be treated to imbue the coatings with better dispersibility, higher brilliance and greater wash-resistance.

Additives affect a number of coating properties, including those of film formation, flow, surface smoothness and texture.

Type of silicone additive	Effects possible
Glycol-modified fluids 1) Slight floating	Flow improvers to prevent floating (non-uniform distribution of pigments)
2) Extensive floating	Improvement in gloss and surface smoothness (scratch resistance)
Alkyl-modified fluids	Effective against flow trouble (silicone pest), additives for primers (can coating), effective flow improvers for curtain coating (wood finishes)
Low-viscosity methyl silicone fluids	Flow improvers
High-viscosity methyl silicone fluids	Hammer effect finishes
Modified siloxanes with acrylic groups	Flow improvers and crosslinkers for radiation-curable acrylic paints (for metal, paper and plastics)

In addition to improving flow properties, silicone fluids facilitate the grinding of pigments during paint manufacture.

19. Electrical and electronics industries

The whole spectrum of silicone products is represented in the electrical and electronics industries, from silicone fluids, all kinds of rubbers right up to silicone resins. HTV rubber, in particular, has proved indispensable for cable insulation.

19.1. Cables and cable accessories

The most important products made by the cable industry are shown below.

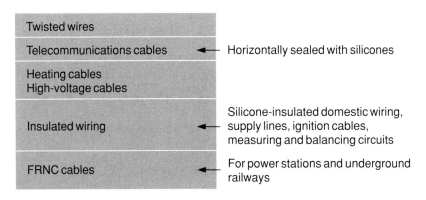

Twisted wires	
Telecommunications cables	← Horizontally sealed with silicones
Heating cables High-voltage cables	
Insulated wiring	← Silicone-insulated domestic wiring, supply lines, ignition cables, measuring and balancing circuits
FRNC cables	← For power stations and underground railways

The primary uses of silicones as cable-insulation material are in telecommunications cables (ca. 30 %) and in insulated wiring (ca. 70 %).

Several prominent properties lean heavily in silicone rubber's favour, especially

– excellent insulating properties, largely regardless of temperature; flame retardancy,

– high resistance to ozone and radiation,

– good rheological properties.

In the following discussion of the various types of silicone-insulated cables, it should be remembered that a distinction is drawn between insulation (for coils) made from silicone resin-reinforced fibreglass or polyester fibreglass and that made from extruded silicone rubber.

Extruded insulation, both with and without serving, is widely used for heating wires, motor junctions, in equipment making, lighting mains and high-voltage cables for television sets.

It ensures that domestic appliances such as irons, grills and powered garden tools function reliably under variable conditions of load and that moisture in washing machines and cold in refrigerators remain ineffectual. Silicone cables are also common in medical devices. Their thermal resistance underpins their use in external illumination, such as halogen lamps. Their high dielectric strength, resistance to corona discharges and thermal resistance make them the articles of choice for television sets and high-voltage appliances. Furthermore, silicone cables ensure that computers, peripheral devices and copying machines function extremely reliably.

Extruded silicone insulation and servings find application as ignition cables in automobiles. The insulating properties are effected with a rubber that is resistant to high tension and that is additionally sheathed in another type which is exceptionally resistant to mechanical stress. The conducting core can further be made of conductive silicone rubber instead of metal.

Multi-core, extruded cables fulfil various low-tension tasks.

Such cables are encountered in instrumentation and control engineering and are resistant to ageing and very reliable (if the cable burns, the silica formed protects the circuit, allowing it to keep functioning). Cables of this type are widely employed in domestic appliances, such as hair dryers, irons etc.

Coaxial cables for telecommunications have to be insulated horizontally with a water-blocking filler to prevent access of water. The filler can be made from petroleum jelly or silicone gel. Telecommunications cables for use off-shore are preferably made from conductive silicone rubber, which allows defects to be localized readily.

Silicone rubber for insulating high-tension multi-core cables is sheathed additionally in braided fibreglass, braided metal and an additional plastic sheathing.

These types of cables provide electricity supplies, are components of control and signal systems, find application in local and long-distance traffic systems and are used by industry. Fire-proof cables also belong to this group. They are insulated with extremely flame-retardant silicone rubber that is further sheathed in mica-fibreglass tape. For reasons of safety, fire-proof cables are preferred in ships (safety and alarm cables), in tunnels, underground railways, gas-powered equipment, ovens and in power stations.

Cable bushings are also insulated with flame-retardant silicone rubber foam.

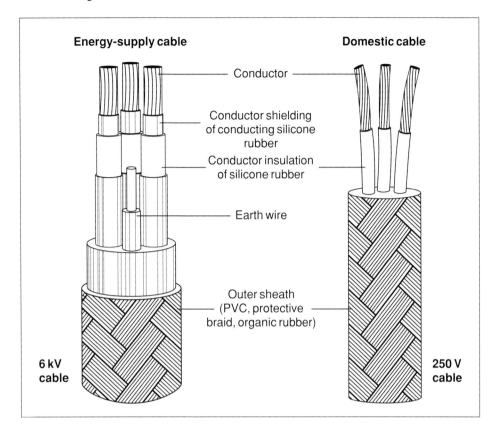

Conductive silicone rubber has special importance in cable technology for extrusion of cable cores. Appropriate smoothing of the conductor prevents corona discharges and raises the cable's dielectric strength.

Silicones are also required for optical waveguides. When glass fibres are being made by the glass-drawing process, they are immediately coated with silicone rubber to protect them against mechanical damage. They are then fed into a plastic tube containing an insulating liquid.

Silicones play other major roles as insulating materials in the cable industry.

The electrical properties of PVC, by far the most popular insulation material (over 50 %), are increased by surface-treating the fillers.

Polyethylene is finding increasing usage as an insulating material (currently around 30 %). This is due primarily to its being post-cured with silanes to XLPE. XLPE has much better electrical properties, particularly in regard to reduced water treeing and electrical treeing.

The cable industry is not alone in using large amounts of paper for insulation purposes (often alternating with polypropylene layers). Such materials are frequently impregnated with insulating liquids (including silicone fluids) in order to enhance their dielectric strength.

Cable accessories are also produced by the cable industry, e.g. sleeves and cable terminations. The latter must

- control the electrical field at the cable end,

- protect the cable core against atmospheric influences,

- seal the cable against moisture and aggressive chemicals, such as airborne pollutants, under fluctuating thermal conditions,

- provide a service life of approximately 30 years,

- be easy to install.

The electrical field at the cable termination is controlled by means of an electrically conducting field-control element. A distinction is drawn between field control effected by deflectors (employing electrically conducting silicone rubber with a conductivity of $25-200\,\Omega\,cm$) and by refractive field control with the aid of a cylindrical component with a resistivity of $10^7-10^9\,\Omega\,cm$ and dielectric constant greater than 150. These techniques ensure approximately linear decay of potential along the core insulation.

Cable terminations featuring capacitive, deflector field control are either of the narrow type or the dish type, according to whether they are located indoors or outdoors.

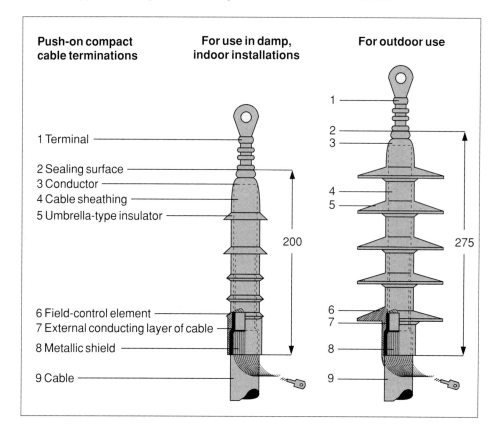

Push-on compact cable terminations

For use in damp, indoor installations

For outdoor use

1 Terminal
2 Sealing surface
3 Conductor
4 Cable sheathing
5 Umbrella-type insulator
6 Field-control element
7 External conducting layer of cable
8 Metallic shield
9 Cable

It has become necessary to use push-on flexible cable terminations since oil-impregnated papers are increasingly being superseded by plastic insulation, especially XLPE, for the medium-voltage range. They allow greater conductor temperatures, which entail great thermal expansion on the part of the insulation.

A push-on cable termination must therefore not suffer any loss in flexibility in the long term and must remain firmly attached to the cable.

Properties which render silicone rubber an ideal material for cable terminations are

– high permanent elasticity and low compression set, and

– tracking resistance under the influence of concentrated electrical discharges.

The tracking resistance of EPDM can only be rendered comparable to that of silicone rubber by adding a high proportion of aluminium oxide trihydrate filler. However, this would increase the Shore hardness disproportionately.

19.2. Electrical insulation materials

In electrical engineering, the insulating materials can be liquids, powders and solids. A breakdown of the various categories is shown on page 144.

For construction purposes in electrical engineering, silicone resins and rubbers are best suited for

– rigid fibreglass and mica laminates, as well as

– flexible fibreglass and mica/fibreglass sheets.

The laminates have a sandwich construction in which mica paper or fibreglass fabric is impregnated with silicone resins and then compressed in the presence of heat by the suspension and solvent processes. The most important application for mica laminates is that of heating micanites for toasters. Adhesive tapes containing a silicone adhesive resin have been developed for high-temperature applications.

Porcelain is primarily employed for outdoor insulators. Treatment with silicone pastes prevents flash-over from occurring in polluted industrial atmospheres.

Other techniques for insulating coils utilize mica papers with subsequent vacuum-pressure impregnation (VPI). It is also possible to use mica papers that have a high resin content and that are only cured under pressure. This is known as the resin-rich technique.

Motors that have been encapsulated with organic resins (usually epoxies) can be further protected against damp by impregnation with silicone rubber dispersions and can be used in the tropics.

The greatest advantage of silicone-encapsulated motors lies in their enhanced resistance to thermal stress. Equipment can be made very compact and savings made on weight. Silicone-insulated motors are preferred in roll mills, traction motors, machine tools and in the chemicals industry.

Common classes of insulating materials, temperature limits, and examples of possible materials (as given in German Standard VDE 0530)

Class	Limiting temperature	Insulating material	Binder, impregnating agent or coating for the manufacture of substances in the next column	Impregnating agent for the finished winding and insulation systems
E	120°C	Cotton/foam laminates Paper/foam laminates	Phenol-formaldehyde resins	Synthetic resin coatings, including oil-modified types; crosslinked polyester resins, epoxy resins
		Treated textiles	Synthetic resin paints, including oil-modified types	
B	130°C	Coated fibreglass textiles	Synthetic resin paints	
		Films based on polyethylene glycol-terephthalate Fabrics or nonwovens made from fibres of polyethylene glycol terephthalate Films based on crystallized polycarbonate	None	Synthetic resin paints, including oil-modified types Crosslinked polyester resins, epoxy resins
F	155°C	Coated fibreglass textiles Mica products with or without carrier substances Cellulose-free laminates with insulating materials of fibres based on aromatic polyamides Fibreglass laminates	Alkyd resins / Epoxy resins / Crosslinked polyester and polyurethane resins Silicone-alkyd resins	Epoxy resins Crosslinked polyester and polyurethane resins
H	180°C	Coated fibreglass textiles Mica products with or without carrier substances Fibreglass laminates	Silicone resins	Silicone resins
		Insulating substances of fibres based on aromatic polyamides Films based on polyamides	None	
C	Above 180°C	Mica* Porcelain and other ceramics, glass, quartz*	None	None or inorganic binders such as glass or cement
		Treated fibreglass textiles Mica products	Silicone resins with high thermal resistance	Silicone resins with high thermal resistance
		Polytetrafluoroethylene	None	None required

* Maximum operating temperature is limited by physical and technical properties

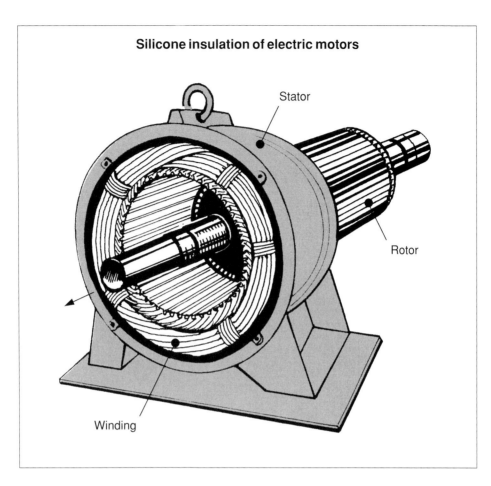

Silicone insulation of electric motors

Rotor	**Armature winding**	Silicone casting resin (VPI)
	Intermediate bearing in vicinity of end windings	Silicone-mica laminates
	Insulation of end windings	Silicone rubber glass-cloth
Stator	**Insulation of winding**	Silicone casting resin (VPI)
	Coil tap insulation	Glass-cloth impregnated with silicone resin

In the field of electrical engineering, a central problem is posed by insulation of tubular heating elements.

In tubular heating elements, the insulating filler, magnesium oxide, has to be protected against moisture absorption. This is accomplished by impregnating the filler with a silicone resin powder or silicone fluid (H-siloxane) and sealing the ends of the element with RTV-1 silicone rubber. The latter application requires rubbers that flow readily and have great thermal resistance.

Tubular heating elements find a myriad of applications in electric cookers, coffee makers, immersion heaters, through-flow heaters, washing machines, dish washers etc.

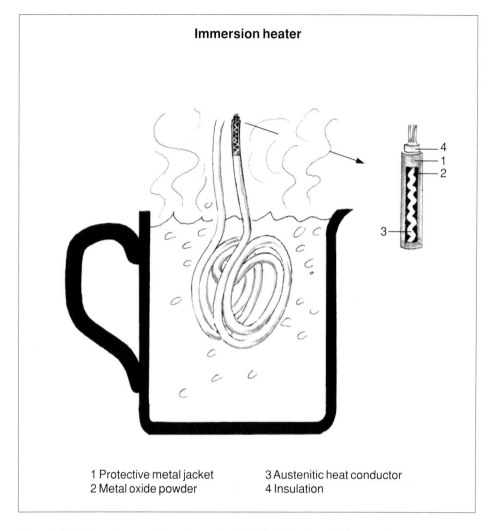

Immersion heater

1 Protective metal jacket
2 Metal oxide powder
3 Austenitic heat conductor
4 Insulation

Electrical devices also contain seals made of HTV rubber (e.g. kitchen appliances, hair dryers etc.). In addition, RTV-1 rubbers are used for caulking (e.g. steam chamber in irons) or bonding (e.g. water reservoir in irons; glass doors in ovens). Microwave ovens contain plates of aluminium that are coated with Teflon® and silicone rubber that has been pigmented with iron oxide.

Conductive silicone rubber is often found in medical devices, e.g. for high-frequency surgery, probes for investigating peristalsis, for electrodes in cephalographs and electro-cardiographs.

19.3. Transformers

Transformers can be divided into dry and liquid types.

In dry transformers, the windings are encapsulated in impregnating resins (cast-resin dry-type transformer); epoxy resins are the most common type in this application. Cast-resin dry-type transformers have great mechanical strength and are used primarily in stationary equipment. A cushion of silicone rubber introduced between the winding and the iron core suppresses acoustic noise.

The situation is somewhat different for the much larger liquid transformers. They contain liquid dielectrics that have to meet the following criteria.

- Outstanding dielectric properties
- Low environmental pollution
- Maximum purity
- Good compatibility with a number of other insulating substances
- Flame retardancy (high flash point and ignition temperature)

Of all the insulating fluids that could be used, ascarels have come under severe attack, because highly poisonous dioxins can form in the event of a fire. Synthetic esters (e.g. of pentaerythritol) have high thermal resistance and are toxicologically and ecologically safe. One drawback is their somewhat high viscosity and susceptibility to hydrolysis (and hence possible change in dielectric properties). Mineral oils are too flammable. Compared with all these products, silicone fluids have very high thermal stability and are toxicologically and ecologically safe.

19.4. Electric heat and lighting technology

Electric heat is based on the principle of the photovoltaic effect, which is utilized in solar cells and allows solar energy to be converted to electric current.

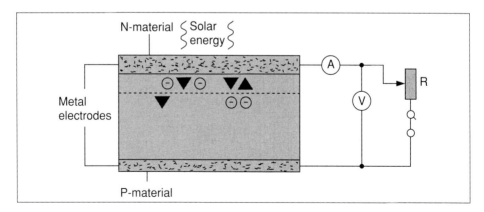

The sensitive solar cells are embedded in silicone rubber to protect them against damage.

In lighting technology, silicone-insulated cables serve as supply lines for street lamps. Lamp-capping cements are based on silicone resin powders and fillers. Incandescent bulbs can be coated with silicone rubber to improve their efficiency.

19.5. Office automation

Modern offices are equipped with photocopying devices, telephones and computers, all of which provide ample scope for using silicones.

Photocopiers operate on the principle of electrostatics in which toner powder becomes electrostatically charged on exposure to light.

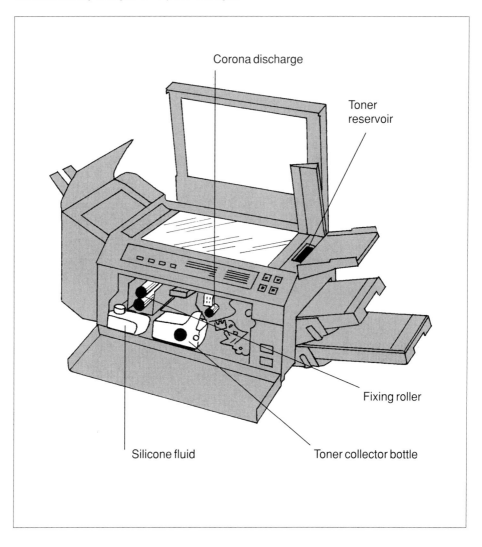

90 % of toner powder consists of magnetic powder, resin binders and fumed silica, which acts as flow improver. Their purpose is to ensure uniform dusting of the powder over the latent (electrostatic) image. To protect the latent image from outside interferences, the feed rollers for the paper are coated with conducting silicone rubber. Once the toner has been applied, it affixes itself permanently to the paper on coming into contact with the thermo-roller.

A number of variants are used to prevent the toner powder from adhering to the rollers. One consists in spraying the rollers with silicone release fluids called fuser oils. In other constructions, the thermo-fixing and feed rollers are coated with silicone rubber. Metal and glass rollers can also be imbued with release properties through treatment with functional silicone fluids.

In the audio industry, silane adhesion promoters help to affix the magnetic powder to the tape.

Conductive silicone rubber has taken over the function of metal springs in keyboards for telephones and similar devices. The usual resistivity of keyboards lies in the range 5 – 10 Ω cm. The logic elements are combined into so-called contact mats, which generally consist of a combination of conductive silicone rubber and a carrier material of highly flexible, insulating silicone rubber.

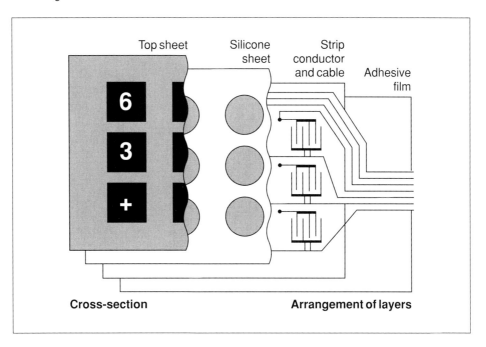

There are various ways of making the contact mats. In one, beads of conductive silicone rubber are vulcanized onto the insulating carrier. In another, the conductive sites on the insulating carriers are applied by screen printing and the use of rubber dispersions.

Contact mats made of conductive and insulating silicone rubber are totally free from bounce and signal delays. Their service life is extremely long, of the order of 40×10^6 contacts. Applications include telephones, remote-control devices for television sets, keyboards for computers and pocket calculators.

The fact that the conductivity of silicone rubber changes under the influence of pressure has been exploited to make pressure sensors (contact resistance in absence of pressure: $> 10^8\,\Omega\,cm$, under pressure: $< 10^2\,\Omega\,cm$). Pressure sensors of this kind are employed in anti-intrusion systems, automatic doors etc.

An important application of silicone rubber is that of electromagnetic interference (EMI) technology. As microelectronics devices become increasingly more widespread, shielding measures have to be taken to prevent interference by electromagnetic fields. On the other hand, electronic devices are being increasingly controlled by ever lower potentials. One consequence of this is cross-coupling, which can cause the devices to malfunction. The most effective way to combat cross-coupling is to shield the devices with seals of conductive silicone rubber (computer seals, rollers of conducting silicone rubber for computer printers).

19.6. Entertainment electronics

There is a lot of scope for using silicones in the field of entertainment electronics, especially in televisions.

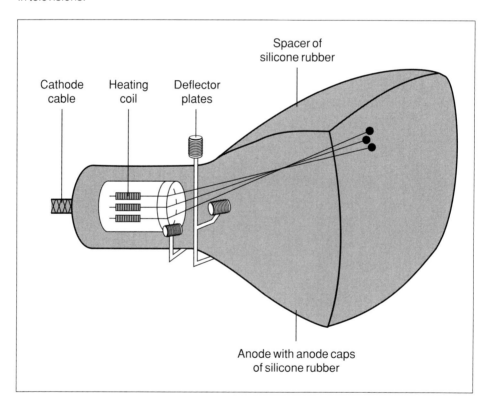

Picture tubes are evacuated with the aid of diffusion pumps, which must be operated with silicone diffusion pump oils in order for the necessary high vacuum to be attained.

The number of silicone parts is rising in televisions, mainly for reasons of safety. For example, the fly-back transformers are increasingly being encapsulated in silicone rubber instead of epoxy resins. Anode caps are a major application of silicones and were originally made of PVC, later of EPDM. Liquid silicone rubber is now favoured. Various spacers are made of silicone rubber to reduce vibration.

In radios, silicone pastes conduct heat away from the power transistors to the heat sink.

19.7. Electronic components

A key operation in electronics consists in creating ultrafine circuits on silicon chips.

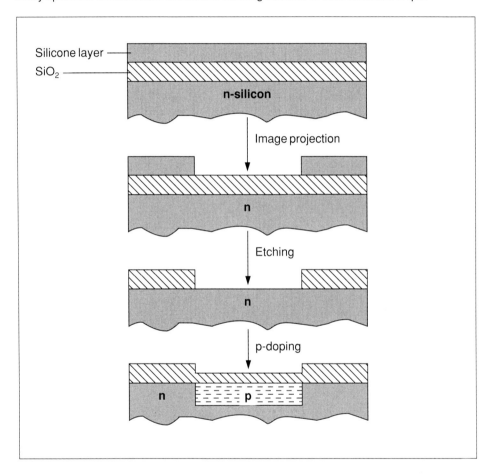

The photolithographic technique depicted above utilizes a so-called photoresist, whose solubility changes on exposure. This means that the exposed resist can be dissolved or etched away. Great hopes have been placed in certain silicon compounds (polysilanes) in this connection. During etching of the substrate surface, sections of the photoresist can become detached. To prevent this from occurring, the oxide surface is treated with hexamethyl disilazane, which reacts with the silanol groups and adsorbed water, yielding a hydrophobic surface to which the photoresist adheres exceptionally well. Quality is thus enhanced and the reject rate diminished.

A further task in electronics is to protect the sensitive components against moisture, mechanical destruction, temperature fluctuations and harmful chemicals.

Of the various dipping and encapsulating materials available, e.g. epoxy resins and polyester resins, silicones excel by virtue of their outstanding electrical properties over a large temperature and frequency range. They can be supplied in a hyperpure form, are readily processed and can be used variously as encapsulating, dipping and adhesive compounds. Curing takes place under mild conditions. Heat is not evolved and shrinkage in volume is negligible.

Chips (ICs) are protected by hermetically sealed ceramic and metal housings and also inexpensive plastic packages. The use of the latter requires additional protection of the encapsulated chip. Epoxy moulding compounds are recommended for smaller chips. Rendering the moulding compound flexible with silicone polymers reduces stress at the substrate interface. However, as chips get progressively larger, the unavoidable shrinkage which accompanies curing of the moulding compound brings with it the danger of detachment and cracking. It therefore becomes necessary to encapsulate the chip further in soft materials.

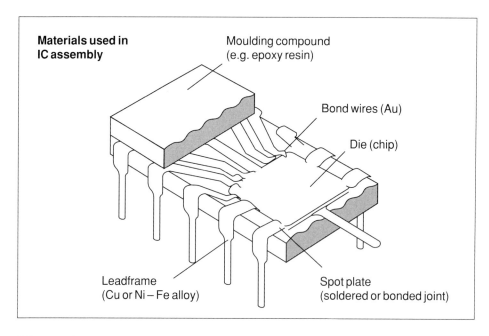

Materials used in IC assembly

Moulding compound (e.g. epoxy resin)

Bond wires (Au)

Die (chip)

Leadframe (Cu or Ni – Fe alloy)

Spot plate (soldered or bonded joint)

Silicone gels and soft silicone rubbers are suitable for this purpose. Owing to their pronounced tack, silicone gels adhere in the absence of a primer, rebond even after forceful detachment, and flow readily. Consequently, there is no risk of damaging the fine bond wires as is the case with moulding compounds. Silicone gels are very good absorbers of vibration and protect components against severe jarring. For this reason, they are almost exclusively used for encapsulating wire-bonded chips on hybrid circuits in automobiles. They afford the best protection against extreme load, water, damp, chemicals, dust and vibration.

Low-viscosity silicone gels are resorted to for encapsulating entire hybrid circuits in conformal coatings. Where individual chips are to be covered, thixotropic silicone gels known as blob tops can be applied. Their thixotropic properties are such that one droplet suffices to completely cover a wire-bonded chip. Apart from covering chips, silicone gels find extensive application for thyristor modules, optoelectronic components (e.g. optocouplers) and solar cells.

Dip-coating of hybrids and circuit boards is more economic in terms of material and money than encapsulating the entire housing. Hard coatings are made from silicone resin whereas soft coatings are made from solvent-free one-pack and two-pack rubbers and low-viscosity UV-curable silicones.

A number of one-pack and two-pack adhesives are available for solving adhesion problems. They are used to bond the housing, to bond the components to aluminium oxide ceramics (thermally conductive silicone adhesives) and to bond chips, e.g. in surface-wave filters (electrically conductive silicone adhesives).

Liquid-crystal displays (LCDs) in calculators and watches are often interconnected with contacts of conductive silicone rubber by what is known as the zebra-pad technique (sandwich structures of conductive and insulating silicone rubber). Liquid crystals have also been developed that are based on siloxanes and cholesterol.

20. Metals, machines, ceramics

Silicone fluids are effective flotation aids in the beneficiation of ores. Steelworks slags containing SiO_2 are employed in various remelting processes for purifying (desulfurizing) steel. Desulfurizing with magnesium is prone to oxidation and hence to combustion but the addition of tiny amounts of silicone fluid to the slag reduces this risk.

In many metal-casting methods, the shell moulds consist of sand bound with synthetic resin (heat-curing phenol resins in the Croning process and cold-curing or "no-bake" formaldehyde-furanol-urea resins). Heat-curing requires the use of silicone release emulsions. Shell strength is enhanced by addition of a silane adhesion promoter to the synthetic resin binder.

The demands imposed by pressure die-casting of aluminium are particularly high. This technique requires special pressure-release emulsions that must not adversely affect coatings applied later. Release agents that are highly compatible with coatings are also used in welding.

Ethyl silicates fulfil important tasks in precision casting. In this technique, the mould material consists of very fine powder (ZrO_2, Al_2O_3) bound together with silicates. Precision casting proceeds by the "lost-wax" or Shaw process and yields extremely accurate castings that require hardly any subsequent machining.

Alloys with low melting-points can be cast direct in silicone rubber moulds. This is known as centrifugal casting and has proved especially suitable for small articles (e.g. jewellery). In electroforming, electrical discharges are used to make metal moulds of the original, with silicone fluids acting as release agents.

Metal-working necessitates the use of cutting oils as coolants and lubricants. Since the oils foam extensively, antifoams are added to the concentrates. Abrasives employed in metal-working contain, for example, SiC as the abrasive grit. Adhesion of the grit to the backing is enhanced by use of silane adhesion promoters.

RTV-1 rubbers are widely used in the construction of machines and equipment. Applications include door seals for refrigerators, seals for pipes etc.

The metal industry affords numerous further opportunities for silicone greases, e.g. lubricating bridge bearings, and for silicone fluids, e.g. lubricating precision instruments.

Copper heat exchangers are coated with silicone-alkyd resin blends. Other typical silicone coatings in the field of process engineering are acid-resistant linings. Flange connectors are made from silicone rubbers.

Ceramics based on inorganic polymers have opened up new construction methods in mechanical engineering.

Polysilanes or polysilazanes yield, after rearrangement to polycarbosilanes, SiC and Si_3N_4 intermediates.

$$\left[\begin{array}{c} CH_3 \\ | \\ -Si- \\ | \\ CH_3 \end{array} \right]_n \xrightarrow{450°C} \left[\begin{array}{c} CH_3 \\ | \\ -Si-CH_2- \\ | \\ H \end{array} \right]_n$$

Compounds of the type SiC and Si_3N_4 possess unusual chemical resistance and high mechanical strength.

One very promising field of application is that of engine design because silicon ceramics can operate at much higher combustion temperatures. This will result in greater fuel economy. Parts already made from silicon ceramics include glow-plug switches for diesel engines, valves and rotors for turbochargers. Likewise, bearings and seals for pumps have been made from silicon ceramics.

Thermal insulators that have to satisfy stringent requirements can be made from consolidated fumed silica. The special network structure of fumed silica greatly reduces thermal conductivity.

21. Transportation

In this field, the automobile offers the most scope for using silicones. The most important applications are shown below.

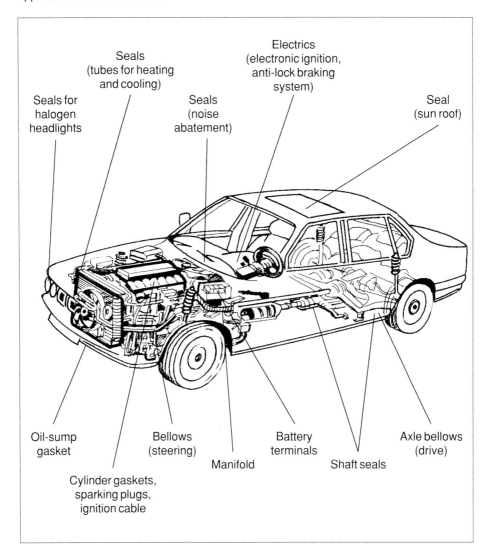

The increasing use of HTV rubber as a working material is due in part to severer operating conditions, such as higher temperatures in the engine compartment, and to major improvements in the quality of silicone rubber, e.g. greater tear strength and oil resistance.

The most common application is that of seals and gaskets for which purpose silicone rubber has superseded other materials such as EPDM and chlorosulfonated PE, 3-ethylene acrylate.

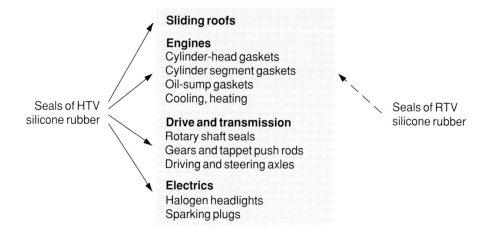

The demands imposed on the various seals differ enormously.

Rotary shaft seals and oil-sump gaskets must be made of rubber that is highly resistant to oil. Furthermore, the rubber for rotary shaft seals must have a high abrasion resistance since they are subject to dynamic load. Diaphragms for the exhaust-return system must be resistant to fuels and their decomposition products as well as being thermostable. Other seals that are severely exposed to fuel are located in the fuel-supply systems.

Seals, caps and diaphragms that only became possible with the development of silicone rubber with enhanced mechanical properties are utilized in drive and steering systems, e.g. front-wheel-drive axles and rack-and-pinion steering.

To withstand the huge amounts of heat evolved by halogen bulbs, headlight caps are made of and sealed with silicone rubber.
The outstanding weathering resistance of silicone rubber is beneficial in seals for sliding sunroofs. Its high resistance to low temperatures renders it a suitable material for diaphragms used in central-locking systems.

Multifarious seals are used in engine blocks, e.g. flat seals (asbestos impregnated with silicone or coated with silicone rubber), RTV-1 seals (formed-in-place gaskets), HTV seals (engine seals and cylinder-head gaskets).

Liquid silicone rubber is increasingly being used for small parts in vehicles on account of the ease with which it can be processed. Examples are pressure-compensation and ventilation valves, multiple plugs in the engine compartment. It is already used to make sparking plugs.

The growing interest in safety and reliability has led to the use of silicone rubber hoses reinforced with a fabric backstay for the heating and cooling system. Manifolds of silicone rubber are installed between the engine and the exhaust system, especially in turbocharged engines. RTV-1 rubber acts as an optimum seal where the radiator is connected to the radiator tank, the radiator lid, intake pipe etc.
Silicone rubber is very efficient at damping acoustic and mechanical vibrations, and seals made from it are highly effective at silencing engine noise.

The Menasco shock-absorber system utilizes silicone polymers. The polymer is forced through a perforated disc when impact occurs and thus absorbs the kinetic energy.

High-polymer silicone fluids are excellent power-transmission media.

In visco clutches, they transmit the power and replace the central differential. Unlike the case with friction clutches, the torque is transmitted continuously as a function of the difference between the rotational speed of the drive and driven shafts.

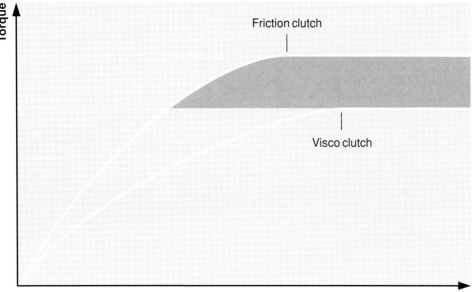

The visco clutch can fulfil various tasks in automobiles. Apart from functioning as a four-wheel-drive distributor differential it can act as a limited-slip differential.

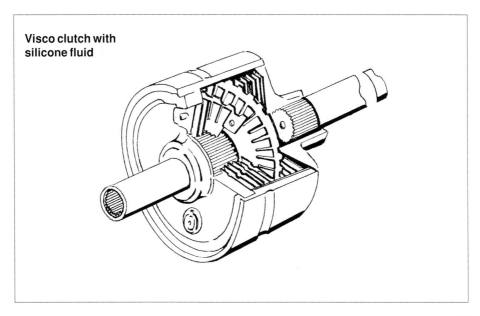

Visco clutch with silicone fluid

Other applications of silicone fluids as power-transmission media are to be found in fan clutches and pendulum dampers.

In the electrical system, the most important application of silicone rubber after sparking plugs is that of ignition cables.

Weather packs for plug-and-socket connections are made from easy-flow, oil-exuding silicone rubber.

Electronics are playing an ever greater role in automobiles especially in ignition control and in anti-lock braking systems. The electronic components are protected with a conformal coating and the devices are fixed with thermally conductive RTV-1 rubbers.

Silicones are used for the same purposes in other means of transport as they are in automobiles.

Airplanes are increasingly being fitted with seals made of rubber that is flame retardant, and resistant to fuel and oil, especially where fuel intake is concerned. Silicone rubber profiles are used to seal windows against the low outside temperatures.

Shipbuilding offers great scope for flame-retardant cables, weathering-resistant paints and caulking compounds made of silicone rubbers. Silicone rubber dispersions have proved to be very good smooth caulking compounds.

In the space industry, silicone resins are used as so-called ablation materials.

22. Foodstuffs, pharmaceuticals and medicine

Water is a prerequisite for preparing foodstuffs. In arid areas, sea-water desalination plants for rendering water potable employ silicone defoamers.

Many basic materials for foodstuffs and pharmaceutical products are obtained via biotechnological processes. They include L-glutamine, vitamins and ethanol for foodstuffs and 6-APA for pharmaceutical raw materials. All biotechnological processes require great quantities of air and intimate mixing of the reaction medium. This, of course, gives rise to foaming problems, caused by such surface-active substances as proteins.

The individual stages in biotechnological processes are enzyme-catalysed fermentation, extraction and isolation of the cells of microorganisms.

Production of material in the fermenters is accompanied by a growth in mass of microorganisms and entails strict adherence to the optimum processing conditions. There are basically three types of biotechnological processes.

– **Change in state of products**
   The products made by the microorganisms alter the state of the substrates. For example, in bread-making, the enzymes in the yeast convert the starch to alcohol and carbon dioxide, which imparts a light texture to the dough during baking.

– **Production of fermented products**
A characteristic feature of these processes is the generation of new products by microorganisms. One age-old process is the conversion of carbohydrates; the formation of alcohol or yeast and carbon dioxide depends on the type of nutrient medium and whether fermentation is carried out aerobically or anaerobically. Other processes are used to prepare citric acid and monosodium glutamate. Silicone defoamers have proved especially invaluable in the latter process.

– **Recovery of bioproducts in cells**
In this kind of biotechnological process, the bioproducts are generated in the cells of the microorganisms themselves. It is crucial for the nutrient broth to be well aerated, a fact which imposes special demands on the antifoam. This process is used to manufacture washing powder enzymes, glucose isomerase, penicillin and inoculants.

When fermentation is complete, the reaction mixture must be processed further. The bioproducts made by the microorganisms are either localized in the cells or are emitted by the cells to the surrounding liquid. This forms the basis for the extraction methods shown in the following diagram.

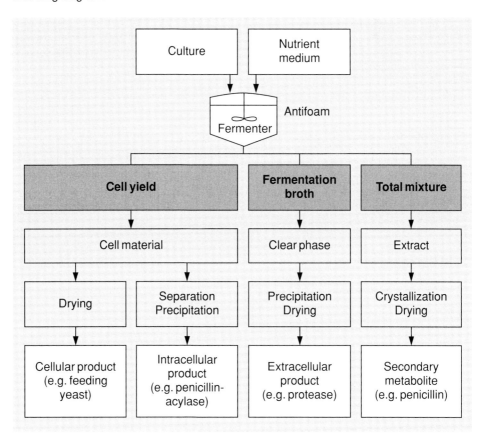

The clear phase indicated in the diagram is removed or processed further and recycled. After washing and renewed drying, cell lysis and separation of cell fragments are performed.

Biotechnological processes have benefited greatly from the immobilization of enzymes (solid-phase catalysis). This technique stabilizes the enzymes for long periods of time and, once the reaction is complete, allows them to be separated readily from the reaction mixture. Enzymes can be immobilized with the aid of silane adhesion promoters:

$$-O-Si(CH_2)_3SH \xrightarrow[\text{Cysteine-enzyme}]{Na_2SO_4} -O-Si(CH_2)_3-S-S-\text{Cysteine-enzyme}$$

One example of solid-phase catalysis is the immobilization of penicillase on eupergit, an epoxy polyacrylate, during the cleavage of side chains in penicillins produced by fermentation. Another is the fixing of lipases to silica gel substrates with aminopropyl silane in the manufacture of glutardinaldehyde. Extensive use is made of solid-phase catalysis for processing foodstuffs and beverages, resolving racemic mixtures, isomerizing fructose, liquefying starch and decomposing yeast.

In addition to being used in the above-mentioned biotechnological processes, silicone products are employed by the foodstuffs sector itself.

For example, iron sulfate solutions containing admixed silicone surfactants are applied to fruit in order to improve intake of trace elements. Oranges are coated with wax to impart a sheen to the surface and to prevent them from drying out. The wax coatings contain silicones that enhance the sheen and prevent foaming from occurring during coating.

Tins and trays in the baking industry are coated with silicones to stop the wares from sticking. Bottle corks are lubricated with silicone fluid emulsions. Beverage-dispensing machines and coffee percolators are fitted with seals of silicone rubber. Tubes and nipples of silicone rubber do not have a tendency to form cracks in which proteins could be deposited and bacterial growth therefore promoted. For this reason, babies' dummies and bottle teats are made of silicone rubber.

Applications in medicine are manifold.

Silicone rubber is used to make all kinds of prostheses, ranging from heart valves, ankle prostheses, artificial joints, mammary prostheses, facial epitheses (tissue substitute), to endoprostheses (bone substitute).

Silicones are used for tubes and catheters for functional organ support, in surgical applications (e.g. for heart-lung machines and blood-transfusions) and as encapsulants for heart pacemakers. Tendons heal easier in silicone rubber veins, which prevent them from adhering during healing. Silicone fluids are very good lubricants for rheumatoid-arthritic joints.

Contact lenses made of hard siloxane alkyd methacrylate have the advantage of being more permeable to oxygen.

In the nursing sector, face masks and respiratory equipment are made of silicone rubber. Bed sores are prevented by silicone fluids.

A major area is that of dental impression compounds in stomatology. Unlike organic impression compounds of polysulfides and alginates, silicone rubber is non-toxic, easy to handle and yields accurate moulds. Silane adhesion promoters are used in dental prostheses.

Equipment for administering drugs, e.g. syringes, commonly contains seals and bungs of silicone rubber.

Silicone rubber also plays a part in transdermal therapeutic systems for the controlled and selective release of drugs. From its rubber matrix, the drug passes into the greater circulation direct, thereby escaping first-pass loss in the liver.

Dispensers made of expanded polyamide coated with silicone rubber have been developed for metering progesterone to synchronize menstruation cycles.

Silicone resins protect tablets against moisture whereas fumed silica causes tablets to disintegrate readily in water.

The last application to be mentioned here is that of silicone antifoams as the active agents in antiflatulence preparations.

Terms used in chemistry and industry

Acids are chemical compounds whose hydrogen atoms can be replaced by metal atoms on reaction with **bases** to form salt and water. This process is known as neutralization.

$$HCl + NaOH \longrightarrow NaCl + H_2O$$

The salts generated are present in the water as cations (e.g. Na^+) and anions (e.g. Cl^-). The acidity or alkalinity of a compound is given by its pH, which is the negative decadal logarithm of the concentration of hydrogen ions.

Adhesives are high-molecular substances that, owing to their property of sticking to surfaces (adhesion) and their internal strength (cohesion), are used to bond a wide variety of objects. The chief types are solvent-free (hot-melt), and solvent-borne (heat-sealing, pressure-sensitive, contact) and are based on plastics (poly(vinyl acetate) and polyacrylate), rubbers (polybutadiene) and aqueous dispersion adhesives.

Building materials include steel, glass (⟶ silicates), lime-bound materials (lime sandstone), cement-bound materials (cement), hydrothermally cured porous substances (aerated concrete), plastics (PVC) etc. Lime-bound materials harden by carbonation:

$$Ca(OH)_2 + CO_2 \longrightarrow CaCO_3 + H_2O$$

Cement is made of calcium aluminium silicates that cure in water in the absence of carbon dioxide.

Catalysts are substances which, when added in small amounts, accelerate the rate of a reaction without appearing in the final product. Certain complexes of noble metals (platinum, rhodium) are effective catalysts. Other major industrial catalysts are the so-called supported type in which the actual catalyst is supported on a carrier substance with a large surface area (e.g. Ziegler-Natta catalysts).

Cellulose is the chief component of wood and plant fibres. It is a carbohydrate that consists of about 3 000 glucose (i.e. sugar) units linked together. It is the raw material for paper, fibres and such chemical substances as adhesives and coatings.

Chemical reactions: Conversion of chemical elements or compounds into other compounds. A distinction is drawn between homogeneous and heterogeneous reactions according to whether the reagents are in the same or different physical state (→ Direct synthesis). Chemical reactions are influenced by various factors, such as temperature, pressure, concentration etc. and may evolve heat (exothermic reactions) or consume heat (endothermic reactions). Catalysts play a major role in accelerating the rate of reactions (→ Catalysts).

Components are electronic devices for controlling the flow of electrons in diodes, transistors and thyristors. Individual components are integrated to form printed circuit boards (copper-laminated, synthetic resin boards with conductor tracks). Electronic parts are protected by conformal coatings, which are thin, curable polymeric films.

Compression set is the residual decrease in the thickness of an elastomer test specimen, expressed in percent, after the removal of a compressive load applied for a specified length of time.

Cracking is the thermal cleavage of long-chained hydrocarbons into smaller molecules, primarily to recover ethylene and benzene from higher-boiling fractions of crude oil (heavy oil). Cracking is performed in various ways at high temperatures and pressures in the presence of special catalysts. Reforming of benzene serves to produce aromatic compounds from naphthenes with elimination of water. Its purpose is to increase the octane number of fuel.

Crystals are substances whose components (atoms, ions, molecules), in the ideal case, are arranged in regularly recurring crystal patterns known as lattices (e.g. SiO_2 has a tetrahedral lattice). However, owing to vacant sites in the lattice, occupied interstices and the presence of impurities, deviations from the ideal occur (→ Semiconductors). The term crystallinity as applied to linear, high polymers denotes the fraction of ordered crystalline regions. Crystalline regions are also encountered in liquid crystals. These are substances which undergo a transition under the action of heat from crystalline solids to partially crystalline liquid phases (e.g. cholesterol liquid crystals) to isotropic liquids. Colour changes occur at the same time, especially under the influence of an electric field (used in liquid-crystal displays, LCDs).

Damped energy is the absorbed portion of vibrational energy that is converted to heat by means of internal friction (measured in the torsional vibration test). A measure of damping is the mechanical loss factor, tg ∂, which is the ratio of loss in mechanical energy to the recoverable energy.

Direct synthesis (Rochow synthesis): Heterogeneous reaction in which halogenoalkylsilanes are manufactured from silicone powder and gaseous monochloromethane (methyl chloride). Accomplished industrially in fluidized bed reactors in which the silicon powder is swirled up by the monochloromethane.

Distillation is the separation of evaporable substances by virtue of their different boiling points, followed by condensation of the vapours to liquids. In fractional distillation, the individual fractions are trapped separately. The formation of azeotropic mixtures is exploited in azeotropic distillation. Separation is enhanced by means of special fillers of ceramic material, glass or clay (column trays).

Dyes for colouring textiles are mainly vat dyes, mordants, basic (cationic) dyes, acid dyes (for wool), substantive dyes (e.g. coupling dyes for cotton) and disperse dyes (mainly for man-made fibres). Dyes are used for colouring flocks, yarns, piece goods and carpets. Dye baths mainly contain dyes, levelling agents (for better adsorption of the dye) and salts (to increase dye yields). Dyeing is performed at the boil in acidic (for cotton) or alkaline (wool) dye baths.

Elastomers are wide-mesh, crosslinked polymers that feature good elasticity. A measure of elasticity is given by the modulus of elasticity, which is derived from the stress-strain curve as the ratio of stress to strain in N/mm^2/%. Elastic behaviour is further characterized by a high degree of reversibility (compression set) that is only observed in loosely cross-linked molecule chains.

The most important types of elastomer are latex, styrene-butadiene (SBR) and butadiene rubber. Crosslinking in butadiene occurs at the double bonds of the main chains.

$$\cdot \quad \cdot \quad -CH_2CH = CHCH_2 - CH_2CH = CHCH_2 - \quad \cdot \quad \cdot \quad \cdot$$
$$\cdot \quad \cdot \quad CH_2CH = CHCH_2 - CH_2CH = CHCH_2 - \quad \cdot \quad \cdot \quad \cdot$$

Crosslinking, however, can also occur via side chains, as in the case of EPDM (ethylene-propylene-diene-polymethylene rubber).

Esters are formed by the reaction of alcohols with acids and have the following formula:

$$RC\underset{OR}{\overset{O}{\diagup\!\!\!\!\diagdown}}$$

Examples of esters are fats, the products of reactions of higher carboxylic acids such as stearic acid with glycerol. Saponification ($\longrightarrow$ Hydrolysis) of fats yields soaps (salts of fatty acids). Enzymes capable of breaking down esters are known as esterases (e.g. lipases).

Fibres are filaments of different length (e.g. endless fibres, staple fibres). According to their chemical constitution, a distinction is drawn between cellulosic fibres (cotton), proteinaceous fibres (wool), man-made fibres (polyester, polyamide, acrylic) and inorganic fibres (asbestos, glass, metal, carbon). Various lubricants are needed to help process them (e.g. sizes, textile preparations, softeners, spinning oils).

Fillers are powders that are added to plastics, coatings, adhesives etc. in order to reduce costs (in the case of chalk and clays) and to impart specific material properties (carbon black, silica). Plastics can be rendered particularly strong by the addition of fibreglass as filler (yielding fibreglass reinforced plastics, especially unsaturated polyester resins). Whiskers are fibrous structures in sinter materials.

Foams: Dispersions of gases in liquids. Foamed materials are products with a porous or cellular structure that have been made with blowing agents, e.g. expanded plastic (Styropor®), foamed building materials (aerated concrete), and foamed artificial leather. Polyurethanes are foamed by chemical reaction of the isocyanate group with water. The CO_2 evolved acts as the blowing agent.

$$R - NCO + H_2O \longrightarrow R - NH_2 + CO_2\uparrow$$

Hydrocarbons are compounds that contain only carbon and hydrogen. Examples include methane, CH_4; ethylene, $CH_2 = CH_2$ (⟶ Plastics); butadiene, $H_2C = CHCH = CH_2$ (⟶ Elastomers). Hydrocarbons are mainly obtained from crude oil (⟶ Cracking).

Hydrolysis: Conversion or decomposition of chemical compounds by water.

Insulators are substances that exhibit high resistance to the passage of an electric current in accordance with the equation:

$$\text{Resistance} = \frac{\text{Voltage}}{\text{Current}}$$

A measure of the insulating action is the dielectric strength (in kV) of an insulator located between two electrodes. The dielectric constant is the ratio of the values obtained for an insulator relative to a vacuum. The dielectric warming experienced by an insulator in an alternating electrical field is given by the dielectric loss factor (tg ∂ is the ratio of the active current to the perpendicular reactive current). Substances with a high tg ∂, such as PVC, can thus be readily heated by high-frequency devices (use in HF welding). Substances with a low tg ∂, such as PE, are good insulators in high-frequency technology.

Oxidation is the uptake of oxygen or, in the extended sense, the donation of electrons and increase in the oxidation number of an element, e.g.,

$$4\ Fe^{(0)} + 3\ O_2 \longrightarrow 2\ Fe_2^{(+3)}O_3$$

An important oxidation series in organic chemistry proceeds from alcohols (e.g. ethyl alcohol, C_2H_5OH) to aldehydes (e.g. acetaldehyde, CH_3CHO) to carboxylic acids (e.g. acetic acid, CH_3COOH). Cleavage of hydrogen from organic compounds is also an oxidation process (⟶ Cracking). The reverse process is known as reduction. The ageing properties of plastics are mainly governed by resistance to oxidation.

Photochemical processes are initiated by the absorption of short-wave radiation, such as UV and X-rays. Ozone formation is a photochemical process. Photoresists are substances that, on exposure to light, undergo changes, e.g. polymerization, chemical degradation, altered solubility. They are used in polygraphy to manufacture microelectronic components and in surface coatings (⟶ Printing techniques).

Plastics are, in the widest sense, all organic macromolecular materials from which consumer articles are made by compression, extrusion and related techniques. A distinction is made between thermoplastics, such as polyolefins, polyamides, PVC, and thermosets, such as phenol resins, epoxy resins. Thermoplastics are macroporously crosslinked, primarily linear polymers that soften reversibly at elevated temperatures. By contrast, thermosetting plastics harden to form extensively crosslinked, lattice polymers that cannot be reshaped under the action of heat (glass-transition temperatures lie above 50°C).

Polyesters: Polymers containing ester groups, particularly resins, which are obtained by esterification of polybasic acids (e.g. phthalic acid, $C_6H_4(COOH)_2$) and polyhydric alcohols (e.g. ethylene glycol [1,2-ethanediol]). In the case of unsaturated polyester resins (abbreviated to UP), at least one component contains a double bond (e.g. maleic acid). Admixing of monomeric, polymerizable compounds such as styrene results in crosslinking at room temperature (hence use as casting resins and coatings). Alkyd resin paints are unsaturated polyester resins that have been modified with fatty acids.

Polyglycols: Primarily polymers of ethylene glycol (1,2-ethanediol, CH_2OH-CH_2OH) and propylene glycol (1,2-propanediol) with the general structure:

$$HO(CH-CH_2O)_nH$$
$$|$$
$$H \text{ or } CH_3$$

Polyethylene glycols are made by the adding-on of ethylene oxide (1,2-epoxyethylene) and water. Adding-on of ethylene oxide to such compounds as alcohols, amines and acids is known as **ethoxylation** and produces ethers, esters or amines of polyglycols of different degrees of ethoxylation (number of attached ethoxy groups). Polyglycols play an important role in the manufacture of surfactants.

Polymers are substances that are "built up" or polymerized from such building blocks (monomers) of low molecular weight as ethylene, vinyl acetate, and vinyl chloride. An example of a polymer is poly(vinyl chloride):

$$-CH_2-CH{\Big[}CH_2-CH{\Big]}CH_2-CH-$$
$$\quad\ \ |\qquad\quad\ \ |\quad\ _n\quad\ \ |$$
$$\quad\ \ Cl\qquad\quad Cl\qquad\quad Cl$$

Polymers are obtained by emulsion, suspension and bulk polymerization, polycondensation and polyaddition. Polymerization can be initiated by catalysts (referred to as initiators), light, radiation and heat. In free-radical polymerization, the polymer chain grows via the formation of free radicals whereas, in ionic polymerization (basic or acidic catalysts used), it grows by means of electric charge. The degree of polymerization is the number of monomers contained in the polymer. If trifunctional molecules are used, the chains can be crosslinked, e.g. during polyaddition (⟶ Polyurethanes). Stereospecific polymerization leads to isotactic or syndiotactic polymers.

Polyurethanes: Polymers obtained by polyaddition of diisocyanates, e.g. $O=C=N(CH_2)_6N=C=O$, and polyols, e.g. butylene glycol (2,3-butanediol, $HO(CH_2)_4OH$. Their general formula is:

$$-NHCOO(CH_2)_4OCONH(CH_2)_6NHCOO-$$

They can be partially crosslinked by trifunctional components and find numerous applications as thermoplastics (→ Elastomers, Fibres, Adhesives, Foams).

Printing techniques: Various reproduction processes that differ according to the type of printing form, such as relief (flexographic), intaglio (gravure), planographic (offset) and screen printing. In electrophotographic techniques, impinging light charges semiconductors (e.g. zinc oxide) electrostatically, causing them to attract certain toner powders.

Proteins: Albuminous substances consisting of amino acids.

Release agents are products that prevent materials from adhering to one another. They are used in the form of coating agents, as release carrier materials and release additives. The term controlled-release effect covers release papers with specific release force values, and drugs and fragrances that release their active substances slowly.

Semiconductors are crystalline solids whose conductivity lies midway between that of metals and non-metals. Examples are silicon, gallium arsenide, zinc oxide and graphite. Their conductivity depends greatly on the concentration of impurities and the number of defects in the crystal lattice and is considerably enhanced by doping with atoms of other elements. A distinction is drawn between n-types (conduction effected by movement of electrons, e.g. silicon doped with phosphorus) and p-types (conduction effected by movement of holes, e.g. silicon doped with boron). p- and n- regions on chips are made by photolithographic means (→ Photochemical processes).

Strength: Parameters determined by materials testing that describe the load which leads to fracture (e.g. compressive strength, tear strength [N/mm^2], tear-propagation strength). Impact strength is the work required to break the test specimen (in J/m^2).

Surface coating compounds are liquids that leave a film on a surface after drying. They consist of binders, pigments, fillers, solvents and additives (e.g. siccatives). Lacquers and varnishes are special coatings that are characterized by toughness, gloss and resistance to weathering.

Textiles are materials that are made of interwoven warp and weft threads in the case of fabrics or of parallel threads in the case of knitted goods. Weaves can be of the linen, twill or atlas type. Textiles can be built up in several layers and embrace foundation, staple and pile (e.g. velvet) threads. The layers in textile non-wovens are attached by bonding or needle-felted. "Bondings" are made by laminating textiles and foamed films together.

Toxicity is the degree of strength of a poison. A distinction is drawn between acute toxicity (absorption or intake of an active agent just once; characterized by LD_{50} values) and subchronic and chronic toxicity (protracted absorption or intake). Effects manifest themselves in the metabolism (conversion of substances produced by the body, e.g. hormones, and from those ingested, e.g. drugs) and in possible initiation of mutations (spontaneous changes in genes and chromosomes).

Index

A

Addition crosslinking 46, 47, 48
Addition reaction 18, 25
Adhesion 59
Alkoxylation 24

B

Batch process 24
β-Lactam antibiotics 81
Boiling point of methyl chlorosilanes 17

C

Calendering 45
Catalyst, condensation 22
Catalyst, copper 13, 14
Catalyst, platinum 47
Catalysts, polymerization 80
Catalyst, tin 46
Ceramics 153
Chain stoppers 22, 23
Compounding of silicone rubber 62, 121
Compression set 50, 52
Copolymers, silicone polyether 35, 36
Corrosion 138
Crepe hardening 45
Crosslinking 45, 46, 47, 94, 95, 119, 121
Crosslinking agents 46, 47, 64
Crosslinking, condensation 47
Crosslinking density 46
Crosslinking, peroxide 48
Crosslinking, radiation 48, 113

D

Damping properties 61, 156
Defoamers 37, 38, 79, 107
Degradation 28, 32, 52

Depolymerization 48
Dielectric properties 30
Dielectric strength 30, 56
Dimethyl dichlorosilane 17
Dimethyl polysiloxanes
– acetoxyalkyl 120
– aminoalkyl 37, 93, 120
– cyclic and linear 20
– hydroxyalkyl 20, 21, 22, 37
Direct synthesis 13
Disilane cleavage 84
Disproportionation 18
Distillation of methylchlorosilanes 16

E

Elastic modulus 50
Electrical properties 56
Electromagnetic fields 143, 151
Emulsion paints 126
Emulsions 40, 41
Equilibration 23
Extrusion 45, 66

F

Fillers 43, 89
Flame retardancy 55
Flash point 33
Fluidized bed reactor 15
Foam, silicone rubber 65
Functionality of siloxane units 13

G

Gas permeability 59
Gels, silicone 64, 152
Glass-transition temperature 28, 52
Grignard synthesis 20

H

Heat, resistance to 28, 29, 33, 52
Hexamethyl disilazane 83
Hexamethyl disiloxane 22
Hydrolysis 20
Hydrosilylation reaction 18

I

Impregnation   104, 125
Injection moulding   45, 66

M

Manufacture
– organofunctional siloxanes   19
– silicone fluids   22, 23
Methanolysis   21
Methyl chloride   13
Methyl chlorosilanes   17
Methyl hydrogen dichlorosilane   17
Methyl hydrogen polysiloxane (H-siloxane)   47
Methyl siliconate   25
Methyl trichlorosilane   17
Molecular weight   35, 42
Mould-making compounds   98, 161

O

Oil, resistance to   54
Organosilicon compounds, flow chart for   26
Ozone, resistance to   52

P

Phenylmethyl polysiloxane   69
Polycondensation   20, 21
Polymerization   22
Polysiloxanes, structure of   12, 13

R

Radiation, resistance to   52
Recombination   18
Relaxation   119
Release agents   42, 100, 101, 111
Release properties   30, 31, 59
Reversion   51, 52
Rheology (flow properties)   60
Rubber, fluorosilicone   69
Rubber, liquid   42, 70, 152
Rubber, rigid   42, 70

S

Seals 99, 157, 158, 159
Selectivity 14, 83
Shrinkage 48, 63, 64
Silanes, recycling of 17
Silanol 20
Silica 26, 43
Silicates 25, 155
Silicon tetrachloride 24
Silylation 81, 82, 83
Sol-Gel process 91
Stabilizers, foam 37, 97
Stress strain behaviour 49, 117
Substitution, nucleophilic 19
Surface activity 32
Surface tension 32
Surfactants, silicone 38, 39, 40

T

Tear-propagation strength 54
Tear strength 54
Tensile strength 53, 54
Tracking resistance 30
Trimethyl chlorosilane 17

V

Vinylmethylpolysiloxane 18, 69
Viscosity 33, 34, 35, 42
Vulcanization rate 63, 65
Vulcanization temperature 70

W

Water repellency 30, 31
Water-repellent treatment
– Building materials 123, 124
– Textiles 104
Water vapour permeability 126, 127, 128
Weathering, resistance to 138

Literature

1. Batzer, H., Polymere Werkstoffe, Georg Thieme Verlag Stuttgart
2. Bock, H., Grundlage der Silicium Chemie, Angewendete Chemie 101 (1989)
3. Deschler, U., Organofunktionelle Silane, Angewendete Chemie, 98 (1986)
4. Flor, S. Miranda, Troubleshooting Guide, Elastomerics, Jan. 1977
5. Hoffmann, W., Technische Elastomere, Kunststoffe, 1987, 10
6. Hoffmann, W., Kautschuktechnologie, Gentner Verlag Stuttgart
7. Houben-Weyl, Methoden der organischen Chemie, Georg Thieme Verlag Stuttgart
8. Lynch, W., Handbook of Silicone Rubber Fabrication, Van Nostrand Reinhold Comp. London
9. Merten, Thermostabile Kunststoffe, Angewendete Chemie 83 (1971)
10. Noll, W., Chemie und Technologie der Silicone, Verlag Chemie
11. Richmond, M.H., β-Lactam Antibiotics, Hoechst AG
12. Timpe, H.J., Photovernetzung von Siliconen, Adhäsion 1985, Heft 10
13. Ullmann, Enzyklopädie der technischen Chemie, Verlag Chemie
14. Voronkov, Silizium und Leben, Akademie Verlag Berlin
15. Wacker leaflet "Brandschutzmittel"
16. Wacker leaflet "Elastische Formen auf RTV-2 Siliconkautschuk für Industrie und Handwerk"
17. Wacker leaflet "Ethylsilikate"
18. Weber, H., Fassadenschutz und Bautensanierung, Expert-Verlag Grafenau
19. West, R., Chemie der Silizium-Silizium-Doppelbindung, Angewandte Chemie, 1987, Heft 12
20. Winacker Küchler, Chemische Technologie, Hanser Verlag
21. Wolfer, D., Elektrisch leitfähiger Silicongummi - ein moderner Werkstoff mit vielen Anwendungsmöglichkeiten, Kautschuk, Gummi, Kunststoffe, Jg. 8 (1981)